TIME

GLOBAL WARMING

Global Warming

EDITOR Kelly Knauer
DESIGNER Ellen Fanning
PICTURE EDITOR Patricia Cadley
WRITER/RESEARCH DIRECTOR Matthew McCann Fenton
COPY EDITOR Bruce Christopher Carr

TIME INC. HOME ENTERTAINMENT
PUBLISHER Richard Fraiman
GENERAL MANAGER Steven Sandonato
EXECUTIVE DIRECTOR, MARKETING SERVICES Carol Pittard
DIRECTOR, RETAIL & SPECIAL SALES Tom Mifsud
DIRECTOR, NEW PRODUCT DEVELOPMENT Peter Harper
ASSISTANT DIRECTOR, BRAND MARKETING Laura Adam
ASSISTANT GENERAL COUNSEL Dasha Smith Dwin
BOOK PRODUCTION MANAGER Jonathan Polsky
MARKETING MANAGER Joy Butts
DESIGN AND PREPRESS MANAGER Anne-Michelle Gallero

SPECIAL THANKS
Bozena Bannett, Alexandra Bliss, Glenn Buonocore, Suzanne Janso, Robert Marasco, Brooke McGuire, Shelley Rescober, Mary Sarro-Waite, Ilene Shreider, Adriana Tierno, Alex Voznesenskiy

A number of the articles in this book were previously published in substantially different form in Time magazine; all have been re-edited and updated. Facts and analysis in this book are current as of July 31, 2007. Time writers and reporters whose work is represented include: "When a Planet Fights a Fever": Writer: Jeffrey Kluger; Reporters: David Bjerklie, Dan Cray, Andrea Dorfman, Greg Fulton, Andrea Gerlin, Rita Healy, Eric Roston. "On the Brink": Writer: David Bjerklie; Reporter: Dan Cray. "The Burden of Asia's Giants": Writer: Bryan Walsh; Reporters: Susan Jakes and Jodi Xu. "When Warming Affects Health": Writer/Reporter: Christine Gorman. "Pioneers" profiles: Writer/Reporters: Charles Alexander, Stefanie Friedhoff, Rita Healy, Hilary Hylton, Michael D. Lemonick, Peter Liu, Krista Mahr, Shoba Narayan, Alex Perry, Eric Roston, Toko Sekiguchi, Bryan Walsh. "Adapting to a Warmer World": Writers: Jeffrey Kluger, Mark Hertsgaard; Reporters: Aryn Baker, David Bjerklie, Dan Cray, Adam Graham-Silverman, Carolyn Sayre, Bryan Walsh. "Building Green": Writer/Reporter: Richard Lacayo. "Who's Cleaning Up Our Act?": Writers: Daren Fonda, Michael Grunwald, Unmesh Kher, Michael D. Lemonick, Margot Roosevelt; Reporters: Steve Barnes, Rita Healy, Adam Pitluk, Ulla Plon, David Thigpen. "Actions": Writer/Reporters: Kathleen Adams, David Bjerklie, Maryanne Murray Buechner, Peter Gumbel, Unmesh Kher, Stephanie Kirchner, Michael D. Lemonick, Laura Locke, Coco Masters, Alice Park, Simon Robinson, Carolyn Sayre, Catherine Sharick, Adam Smith, Bryan Walsh

ISBN 10: 1-933821-23-X
ISBN 13: 978-1-933821-23-8
Library of Congress Control Number: 2007904841

We welcome your comments and suggestions about Time Books. Please write to us at:
Time Books • Attention: Book Editors • PO Box 11016 • Des Moines, IA 50336-1016

If you would like to order any of our hardcover Collector's Edition books, please call us at 1-800-327-6388 (Monday through Friday, 7 a.m.–8 p.m., or Saturday, 7 a.m.–6 p.m., Central time).

PRINTED IN THE UNITED STATES OF AMERICA

PHOTOGRAPHY CREDITS
Front cover: Tui De Roy—Minden Pictures
Back cover, from top: Peter Essick—Aurora; Richard Hamilton Smith—Corbis; Galen Rowell—Corbis; Gideon Mendel—Corbis

The tide is high *An abandoned home in Nags Head on North Carolina's Outer Banks testifies to erosion due to rising sea levels. Beach erosion also forced the 2000 relocation of the nearby Cape Hatteras lighthouse some 2,900 ft. inland from its previous oceanside location*

GARY BRAASCH/EARTHUNDERFIRE.COM

TIME

Contents

Shelved? *The Fimbur Ice
Shelf in Antarctica; the melting
of such vast expanses of ice
due to global warming could
inundate coastal cities*

A Rather Large Question

WHEN HENRY LUCE AND BRITON Hadden founded TIME in 1923, they declared their new magazine would cover the entire world. They probably never imagined that more than 80 years later, the fate of the entire world would be one of the most urgent stories of the day. But that is indeed the case: the issue of global warming promises to be *the* great story of the 21st century, one that will affect every one of us, as well as our children and their children—and one for which such a seemingly portentous phrase as "the fate of the entire world" is for once not a hyperbole but rather a simple statement of fact.

TIME has been reporting on global warming since scientists first began tracing its symptoms. We published our initial cover story on the topic in October 1987. "It is too soon to tell whether unusual global warming has indeed begun," wrote Michael D. Lemonick. But if the climate did begin to change, he reported, we could expect "dramatically altered weather patterns, major shifts of deserts and fertile regions, intensification of tropical storms and a rise in sea level." Sound familiar?

The year after that story appeared, TIME's editors surprised the world—and jump-started a surge of interest in environmental issues—by choosing not to name an individual as their annual Person of the Year; instead, they designated "Endangered Earth" the Planet of the Year. Since that time, the magazine has reported frequently on both broad issues of the environment and the specific topic of global warming.

TIME has now published six cover stories and hundreds of separate reports addressing the steadily worsening climate crisis. An April 2006 cover story made the case that at long last the debate was over, the verdict was in, and the world was irrefutably warming. A year later, a cover story reported on the actions concerned people around the world were taking to both mitigate and adapt to climate change.

This book offers TIME the opportunity to update its reporting on this critical issue, to augment its coverage with fresh pictures and graphics and to offer an all-new briefing on the nuts and bolts of the story. We have deliberately chosen not to focus on the politics of the crisis in this book; those issues are covered in the weekly magazine, and any report we might offer in these pages on that constantly evolving landscape would be immediately dated.

Instead, this book offers readers a solid overview of the subject: the scientific principles behind the warming phenomenon; the symptoms and potential effects of climate change; the essential work being done by researchers, political leaders, architects, engineers and local activists to address the crisis. And the book concludes with a host of basic, hands-on actions each of us can take to ensure that we play a positive role in the story that is going to affect each of us in the years to come: the fate of the entire world.

—*The Editors*

Sea change *The emission of carbon-heavy gases from facilities like this oil rig in the North Sea is a primary contributor to the trapping of solar heat within Earth's atmosphere. That greenhouse effect, in turn, is a principal driver of the gradual warming of the planet*

Problems

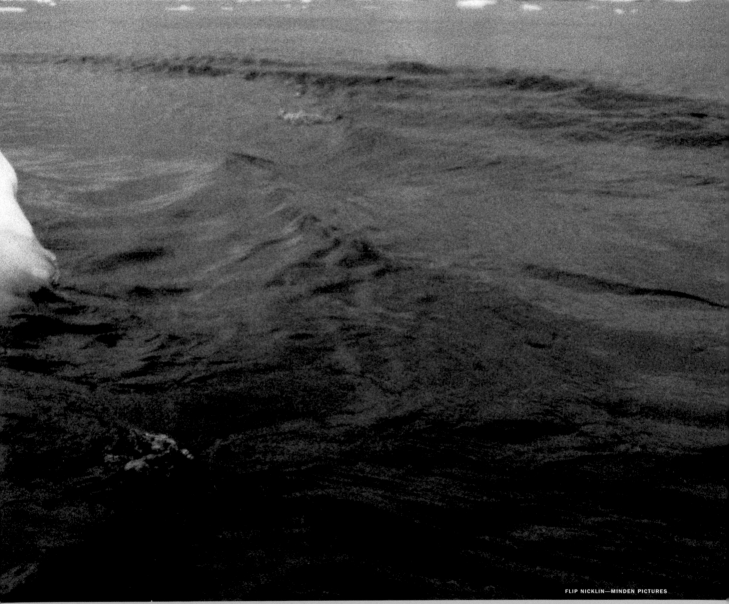

FLIP NICKLIN—MINDEN PICTURES

Like a polar bear clinging to a shrinking ice floe in the Arctic Ocean, many of us have held on to the dwindling hope that global warming is a vague concern for the future. But as extreme weather patterns disrupt lives everywhere, it is clear that climate change is an immediate threat to our planet that must be addressed now

Fueling the Crisis

Hydrocarbon fuels (chiefly coal, oil and natural gas) were first used to create energy at the dawn of the Industrial Revolution, as mankind began to run short of traditional fuels like wood and whale oil. The new fuels contained more latent energy than any previously discovered, but they also came with a flaw: the carbon emissions that are an unavoidable by-product of their combustion (seen here as smoke from a coal plant, in Conesville, Ohio). These gases build up in the planet's atmosphere, where they act like the panes of glass on a greenhouse, allowing the sun's warmth in but not back out. As heat that would ordinarily radiate off into space is instead reflected once again toward the planet's surface, temperatures begin to rise and the globe begins to warm.

PROBLEMS

Burning coal creates power—and releases gases that hel

rive global warming. Can we clean up our act in time?

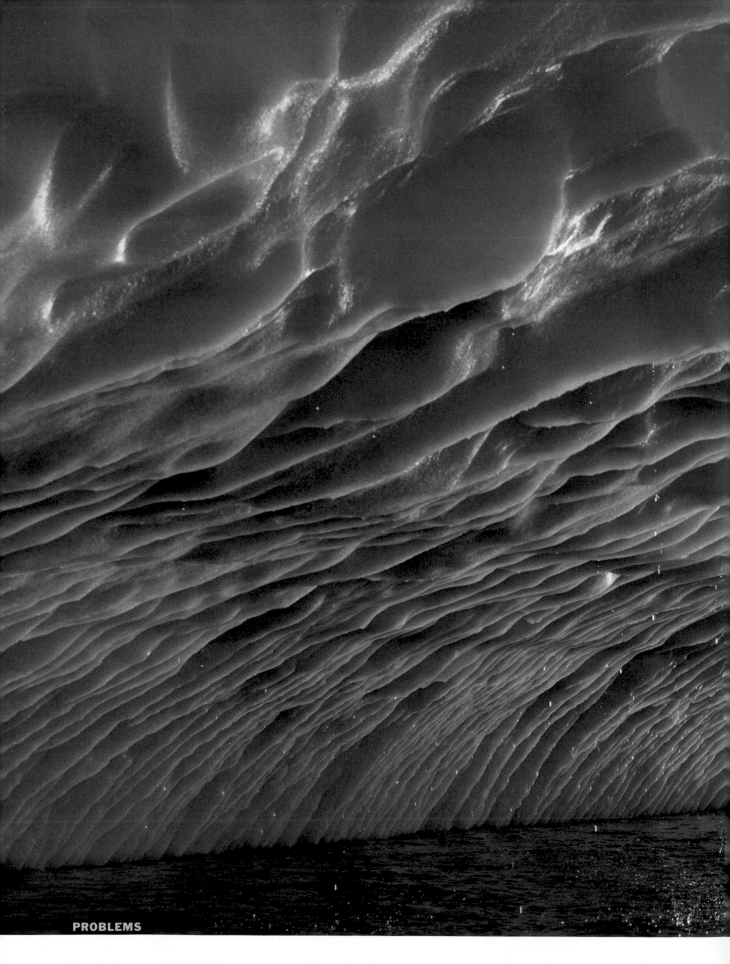

PROBLEMS

As formerly frosty regions heat up, **melting ice** is raising se

A Drop-by-Drop Disaster

As glaciers in Greenland heat up, they are dumping water into the ocean at a rate greater than one cubic mile per week, scientists determined in 2005. For perspective, one cubic mile of water is about five times the amount Los Angeles uses in a year. The concern: that's more than twice the previously observed rate of melt and runoff. By some estimates, if the entire Greenland ice sheet (here turning to slush near Ililussat in 2006) were to melt, it would disgorge enough water to raise global sea levels 23 ft., swallowing up large parts of coastal Florida, most of Bangladesh and many other regions worldwide.

vels, threatening to inundate coastlines around the world

Heat, Floodwaters and Risk Are Rising

Two of the sharpest spears created by global warming—flooding and drought—account for more than half the world's deaths from natural disaster. Nations whose land lies close to sea level, like the Netherlands, Bangladesh and Indonesia, are at great risk, as are locations where natural barriers against water have been degraded by man. Scientists believe that Hurricane Katrina would have packed a far less devastating punch if the wetlands that buffer New Orleans from the Gulf of Mexico had not been removed by imprudent levee-building and oil and gas development. As warming creates extreme weather patterns, lives are disrupted. In February 2007, as many as 400,000 Indonesians in the capital city, Jakarta, shown here, were displaced by huge floods that followed several days of torrential rainstorms.

When severe rainstorms bred **floods** in Indonesia in 2007,

any as **400,000 residents were forced from their homes**

California Conflagration

Wildfires are both a cause and a result of global warming. Parched, arid conditions caused by higher temperatures and less rainfall are nature's recipe for a tinderbox. Add lightning, and you've got a raging wildfire that can consume hundreds of square miles of woodland in a few days. Result: the increasing incidence and intensity of such fires around the world. Here, the Griffith Observatory in Los Angeles is threatened by a wildfire on May 9, 2007.

Fire also helps forge climate change. In the developing world, woodlands and rain forests are being cleared at a staggering rate to make way for crops and livestock grazing, almost always with the use of deliberately set fires. Every second of every day, a slice of forest the size of a football field goes up in smoke, pumping millions of tons of carbon into the atmosphere and taking more of the planet's natural carbon-mitigation engines—trees—out of service. Satellite photos have detected more than 70,000 fires burning simultaneously in the Amazon River rain forest.

In a spiral of flame, vast **fires** are a both the result of global

varming and a trigger that accelerates its pace

Buried Treasure

One paradox of global warming is that it can cause both too much and too little rain, flooding some areas while others suffer prolonged drought. Australia has been parched for more than five years, making the water crisis a central issue in that country's politics. In southwest China's Sichuan region, now enduring its worst drought in more than a century, a 2006 crisis left more than 17,000 people without drinking water. In 2007 the government- and university-operated U.S. Drought Monitor website reported that half the continental U.S. was experiencing abnormal dryness or drought. Over the previous four years, drought patterns migrated east, crippling crops, shrinking lakes and drying up wells in eight Southern states. Florida's Lake Okeechobee, the second largest body of freshwater in America, fell to a record low level in 2007, leaving so much of the lake bed dry that 12,000 acres of it caught fire in May.

Drought can lead to desertification, the process by which once arable land is replaced by arid desert. In some cases, including the current humanitarian crisis in the Darfur region of Sudan, war and genocide are the ghastly offspring of drought, as shrinking supplies of water and fertile land breed violence. Here, farmers in Garissa, Kenya, seek water in 2006 at a well dug by hand.

PROBLEMS

Drought stalks the globe, turning fertile farms into deserts,

arking wildfires, uprooting communities and breeding violence

When a Planet Fights a Fever

The Tipping Point. Climate change isn't science fiction, it's science fact—and it's already damaging the planet at a pace that's alarming. Is it too late to reverse or adapt to the changes?

NO ONE CAN SAY EXACTLY WHAT IT LOOKS LIKE WHEN a planet takes ill, but it probably looks a lot like Earth. Never mind what you've heard about global warming as a slow-motion emergency that would take decades to play out. Suddenly and unexpectedly, the crisis is upon us.

It certainly looked that way in 2007, as some 400,000 Indonesians fled their homes amid rain and flash floods of historic proportions, only two years after curtains of fire and dust turned the nation's skies orange, thanks to drought-fueled blazes that swept across its islands. It certainly looked that way in 2006, as the atmospheric bomb that was Cyclone Larry—a Category 5 storm with wind bursts that reached 180 m.p.h.—exploded through northeastern Australia. It certainly looked

that way as sections of ice the size of small states calve from the disintegrating Arctic and Antarctic glaciers and ice shelves. Weather disasters have always been with us and surely always will be. But when they hit this hard and come this fast—when the emergency becomes commonplace—something has gone grievously wrong. That something is global warming.

The image of Earth as organism—famously dubbed Gaia by environmentalist James Lovelock—has probably been overworked, but that's not to say the planet can't behave like a living thing, and these days, it's a living thing fighting a fever. From heat waves to storms to floods to fires to massive glacial melts, the global climate seems to be crashing around us. Scientists have been calling this shot for decades. This is precisely what they have been warning would happen if we continued

14

Afflictions *The symptoms of global warming are complex, afflicting some with flood and others with drought. At left, rising sea levels due to polar melting threaten residents of Tuvalu in the South Pacific in 2005. Scientists fear the tiny island nation may vanish beneath the waves. Above, farmer John Magill inspects failing crops in Parkes, Australia, in 2006, where a years-long drought is the worst in the nation's history*

pumping greenhouse gases into the atmosphere, trapping the heat that flows from the sun and raising global temperatures.

The U.N.'s Intergovernmental Panel on Climate Change issued a report on the state of planetary warming in February 2007 that was surprising only in its utter lack of hedging. "Warming of the climate system is unequivocal," the report stated. What's more, there is "very high confidence" that human activities since 1750 have played a significant role by overloading the atmosphere with carbon dioxide (CO_2), hence retaining solar heat that would otherwise radiate away. The report concluded that while the long-term solution is to reduce the levels of CO_2 in the atmosphere, for now we're going to have to dig in and prepare, building better levees, moving to higher ground and abandoning vulnerable floodplains altogether.

Environmentalists and lawmakers spent years shouting at one another about whether the grim forecasts were true, but in the past five years or so, the serious debate has quietly ended. Global warming, even most skeptics have concluded, is the real deal, and human activity has been causing it. If there had been any consolation, it was that the glacial pace of nature would give us decades or even centuries to sort out the problem.

But glaciers, it turns out, can move with surprising speed, and so can nature. What few people reckoned on was that global climate systems are booby-trapped with tipping points and feedback loops, thresholds past which the slow creep of environmental decay gives way to sudden and self-perpetuating collapse. Pump enough CO_2 into the sky, and that last part per million of greenhouse gas behaves like the 212th degree Fahr-

enheit that turns a pot of hot water into a plume of steam. Melt enough Greenland ice, and you reach the point at which you're not simply dripping meltwater into the sea but dumping whole glaciers. Studies in 2005 found that several Greenland ice sheets have doubled their rate of slide; in 2006 the journal *Science* published a study suggesting that by the end of the century, the world could be locked in to an eventual rise in sea levels of as much as 20 ft. Nature, it seems, has finally got a bellyful of us.

"Things are happening a lot faster than anyone predicted," Bill Chameides, chief scientist for the advocacy group Environmental Defense and a former professor of atmospheric chemistry, told TIME in 2006. "The last 12 months have been alarming." Says Ruth Curry of the Woods Hole Oceanographic Institution in Massachusetts: "The ripple through the scientific community is palpable."

And it's not just scientists who are taking notice. Even as nature crosses its tipping points, the public seems to have reached its own. For years, popular skepticism about climatological science stood in the way of addressing the problem, but the naysayers—many of whom were on the payroll of energy companies—have become an increasingly marginalized breed. In a 2006 TIME/ABC News/Stanford University poll, 85% of respondents agreed that global warming probably is happening. Moreover, most respondents said they wanted some action taken. Of those polled, 87% believed the government should either encourage or require lowering of power-plant emissions, and 85% thought something should be done to get cars to use less gasoline. Even Evangelical Christians, once among the doubters,

15

Heat wave *China's dry, blistering summer of 2006 drove people in Qingdao to the beach. In some places, temperatures topped 104°F*

are demanding action. In February 2006, 86 Christian leaders formed the Evangelical Climate Initiative and called on Congress to regulate greenhouse gases.

A critical factor in changing the public's view of the climate crisis was former Vice President Al Gore's Powerpoint-driven documentary *An Inconvenient Truth*. Released in May 2006, the film was a surprise smash at the box office, and Gore returned to Washington in March 2007 to testify on global warming to environmental committees in both the Senate and the House.

After a prolonged period of denial, even President George W. Bush, no favorite of greens, now acknowledges climate change and boasts of the steps he is taking to fight it. Before heading to Germany in June 2007 for a G-8 summit meeting, Bush called for a new international accord to fight climate change to be in place by the end of 2008. But most of the measures he supports involve voluntary, not mandatory, emissions controls, rather than the laws with teeth scientists are calling for. At the summit, Bush explicitly rejected mandatory caps on emissions.

Is it too late to reverse the changes global warming has wrought? That's still not clear. Reducing our emissions output year to year is hard enough. Getting it low enough so that the atmosphere can heal is a multigenerational commitment.

Carbon Dioxide and the Poles

AS A TINY COMPONENT OF OUR ATMOSPHERE, CARBON DIOXIDE helps warm Earth to comfort levels we are all used to. But too much of it does an awful lot of damage. The gas represents just a few hundred parts per million (p.p.m.) in the overall air blanket, but they're powerful parts because they allow sunlight to stream in but prevent much of the heat from radiating back out. During the last ice age, the atmosphere's CO_2 concentration

was just 180 p.p.m., putting Earth into a deep freeze. After the glaciers retreated but before the dawn of the modern era, the total had risen to a comfortable 280 p.p.m. In just the past century and a half, we have pushed the level to 381 p.p.m., and we're feeling the effects. Of the 20 hottest years on record, 19 occurred in the 1980s or later. According to climate scientists, 2005 was one of the hottest years in more than a century; so was 2006.

It's at the North and South poles that those steambath conditions are felt particularly acutely, with glaciers and ice caps crumbling to slush. Once the thaw begins, a number of mechanisms kick in to keep it going. Greenland is a vivid example. Late in 2005, glaciologist Eric Rignot of the Jet Propulsion Laboratory in Pasadena, Calif., and Pannir Kanagaratnam, a research assistant professor at the University of Kansas, analyzed data from weather satellites and found that Greenland ice is not just melting but doing so more than twice as fast as in earlier years, with 53 cu. mi. draining away into the sea in 2005 alone, compared with 22 cu. mi. in 1996. A cubic mile of water is about five times the amount Los Angeles uses in a year.

Dumping that much water into the ocean is a very dangerous thing. Icebergs don't raise sea levels when they melt because they're floating, which means they have displaced all the water they're ever going to. But ice on land, like Greenland's, is a different matter. Pour that into oceans that are already rising (because warming water expands), and you run the risk of deluging shorelines. By some estimates, the meltdown of the entire Greenland ice sheet would be enough to raise global sea levels more than 20 ft., swallowing up large parts of coastal Florida and most of Bangladesh. The Antarctic holds enough ice to raise sea levels more than 215 ft.

"Ecosystems are usually able to maintain themselves," says

"Ecosystems are usually able to maintain themselves. But eventually they get pushed to the limit of tolerance"

Terry Chapin, a biologist and professor of ecology at the University of Alaska, Fairbanks. "But eventually they get pushed to the limit of tolerance."

Feedback Loops

ONE OF THE REASONS THE LOSS OF THE PLANET'S ICE COVER IS accelerating is that as the poles' bright white surface shrinks, it changes the relationship of Earth and the sun. Polar ice is so reflective that 90% of the sunlight that strikes it simply bounces back into space, taking much of its energy with it. Ocean water does just the opposite, absorbing 90% of the energy it receives. The more energy it retains, the warmer it gets, with the result that each cubic mile of ice that melts vanishes faster than the mile that preceded it.

That is what scientists call a feedback loop, and it's a nasty one, since once you uncap the Arctic Ocean, you unleash another beast: the comparatively warm layer of water about 600 ft. deep that circulates in and out of the Atlantic. "Remove the ice," says Woods Hole's Curry, "and the water starts talking to the atmosphere, releasing its heat. This is not a good thing."

A similar feedback loop is melting permafrost, usually defined as land that has been continuously frozen for two years or more. There's a lot of earthly real estate that qualifies, and much of it has been frozen much longer than two years: since the end of the last ice age, or at least 8,000 years ago. Sealed inside that cryonic time capsule are layers of partially decayed organic matter, rich in carbon. In high-altitude regions of Alaska, Canada and Siberia, the soil is warming and decomposing, releasing gases that will turn into methane and CO_2. That, in turn, could lead to more warming and permafrost thaw, says research scientist David Lawrence of the National Center for Atmospheric Research (NCAR) in Boulder, Colo. And how much carbon is socked away in Arctic soils? Lawrence puts the figure at 200 gigatons to 800 gigatons. The total human carbon output is only 7 gigatons a year.

One result of all that is warmer oceans, and a result of warmer oceans can be, paradoxically, colder continents within a hotter globe. Ocean currents running between warm and cold regions serve as natural thermoregulators, distributing heat from the equator toward the poles. The Gulf Stream, carrying warmth up from the tropics, is what keeps Europe's climate relatively mild. Whenever Europe is cut off from the Gulf Stream, temperatures plummet. At the end of the last ice age, the warm current was temporarily blocked, and temperatures in Europe fell as much as 10°F, locking the continent in glaciers.

What usually keeps the Gulf Stream running is that warm water is lighter than cold water, so it floats on the surface. As it reaches Europe and releases its heat, the current grows denser and sinks, flowing back to the south and crossing under the northbound Gulf Stream until it reaches the tropics and starts to warm again. The cycle works splendidly, provided the water remains salty enough. But if it becomes diluted by freshwater, the salt concentration drops, and the water gets lighter, idling on top and stalling the current.

Fuming *A traffic jam idles motorists in Bangkok; carbon emissions from gasoline-burning cars are one of the causes of global warming*

From ice to water *These two photographs, taken over a span of 76 years, illustrate the extent of glacial ice melting in South America. At top, the Upsala Glacier in Patagonia, Argentina, photographed in 1928. At bottom, the valley is shown from the same vantage point in 2004. The glacier has receded into the mountains, and a large lake has formed in the valley. Scientists say the glacier is retreating 180 ft. per year*

In December 2005, researchers associated with Britain's National Oceanography Center reported that one component of the system that drives the Gulf Stream has slowed about 30% since 1957. It's the increased release of Arctic and Greenland meltwater that appears to be causing the problem, introducing a gush of freshwater that's overwhelming the natural cycle.

In a warming world, it's unlikely that any amount of cooling that resulted from changing ocean patterns would create a new ice age in Europe, but it could make things very uncomfortable. "The big worry is that the whole climate of Europe will change," says Adrian Luckman, senior lecturer in geography at the University of Wales, Swansea. "We in the U.K. are on the same latitude as Alaska. The reason we can live here is the Gulf Stream."

Drought

AS FAST AS GLOBAL WARMING IS TRANSFORMING THE OCEANS and the ice caps, it's having an even more immediate effect on land. People, animals and plants living in dry, mountainous regions like the western U.S. make it through summer thanks to snowpack that collects on peaks all winter and slowly melts off in warm months. Lately the early arrival of spring and the unusually blistering summers have caused the snowpack to melt too early, so that by the time it's needed, it's largely gone. Climatologist Philip Mote of the University of Washington has compared decades of snowpack levels in Washington, Oregon and California and found that they are a fraction of what they were in the 1940s, and some snowpacks have vanished entirely.

Global warming is tipping other regions of the world into drought in different ways. Higher temperatures bake moisture out of soil faster, causing dry regions that live at the margins to cross the line into full-blown crisis. Meanwhile, El Niño events—the warm pooling of Pacific waters that periodically drives worldwide climate patterns and has been occurring more frequently in global-warming years—further inhibit precipitation in dry areas of Africa and East Asia. According to a recent study by NCAR, the percentage of Earth's surface suffering drought has more than doubled since the 1970s.

Flora and Fauna

HOT, DRY LAND CAN BE MURDER ON PLANTS AND ANIMALS, AND both are taking a bad hit. Wildfires in such regions as Indonesia, the western U.S. and even inland Alaska have been increasing as timberlands and forest floors grow more parched. The blazes create a feedback loop of their own, pouring more carbon into the atmosphere and reducing the number of trees, which absorb CO_2 and release oxygen.

Those forests that don't succumb to fire die in other, slower ways. Connie Millar, a paleoecologist for the U.S. Forest Service, studies the history of vegetation in the Sierra Nevada. Over the past 100 years, she has found, the forests have shifted their tree lines as much as 100 ft. upslope, trying to escape the heat and drought of the lowlands. Such slow-motion evacuation may seem like a sensible strategy, but when you're on a mountain, you can go only so far before you run out of room. "Sometimes we say the trees are going to heaven because they're walking off the mountaintops," Millar says.

The U.S. is home to less than 5% of the world's people, yet it produces 25% of the CO$_2$ emissions on the planet

What About Us?

IT IS FITTING, PERHAPS, THAT AS THE SPECIES CAUSING ALL THE problems, we're suffering the destruction of our habitat too, and we have experienced that loss in terrible ways. Ocean waters have warmed by a full degree Fahrenheit since 1970, and warmer water is like rocket fuel for typhoons and hurricanes. Two 2005 studies found that in the past 35 years the number of Category 4 and 5 hurricanes worldwide has doubled while the wind speed and duration of all hurricanes have jumped 50%.

Since atmospheric heat is not choosy about the water it warms, tropical storms could start turning up in some decidedly nontropical places. "There's a school of thought that sea surface temperatures are warming up toward Canada," says Greg Holland, senior scientist for NCAR in Boulder. "If so, you're likely to get tropical cyclones there, but we honestly don't know."

What We Can Do

SO MUCH ENVIRONMENTAL COLLAPSE HAPPENING IN SO MANY places at once has at last awakened much of the world, particularly the 141 nations that have ratified the Kyoto treaty to reduce carbon emissions—an imperfect accord, to be sure, but an accord all the same. The U.S., however, which is home to less than 5% of Earth's population but produces 25% of CO$_2$ emissions, remains intransigent. Many environmentalists declared the Bush Administration hopeless from the start, and while that may have been premature, it's undeniable that the White

House's environmental record—from the abandonment of Kyoto to the President's broken campaign pledge to control carbon output to the relaxation of emission standards—has been dismal. Faced with such resistance, many environmental groups have resolved simply to wait out this Administration and hope for something better in 2009.

"There are a whole series of things that demonstrate that people want to act and want their government to act," says Fred Krupp, president of Environmental Defense. Krupp and others believe that we should probably accept that it's too late to prevent CO$_2$ concentrations from climbing to 450 p.p.m. (or 70 p.p.m. higher than where they are now). From there, however, they hope to stabilize them and start to dial them back down.

That goal should be attainable. Curbing global warming may be an order of magnitude harder than, say, eradicating smallpox or putting a man on the moon. But is it moral not to try? We did not so much march toward the environmental precipice as drunkenly reel there, snapping at the scientific scolds who told us we had a problem.

The scolds, however, knew what they were talking about. In a solar system crowded with sister worlds that either emerged stillborn like Mercury and Venus or died in infancy like Mars, we're finally coming to appreciate the knife-blade margins within which life can thrive. For more than a century we've been monkeying with those margins. It's long past time we set them right. ∎

Retreat *A drawing on an 1870 postcard shows the Rhone Glacier sweeping into Gletsch, Switzerland; in 2005 the glacier could hardly be seen*

State of The Planet

WARMING TREND

1.44°F
Total temperature rise,
20th century

1.98°F to 11.52°F
Total temperature rise
anticipated in 21st century
if present trends continue

11
Number of the past 12 years
that rank as the warmest in
modern history

SEA LEVELS

4 in. to 8 in.
Global sea-level rise,
20th century

19 in. to 37 in.
Global sea-level rise anticipated
in 21st century if present
trends continue

23 ft.
Additional sea-level rise
anticipated if entire Greenland
ice sheet were to melt

A thin blue envelope of air swathes our planet, and that atmosphere makes possible life as we know it. Now that essential, invisible barrier is loading up with gases that help trap solar heat, warming the planet's surface. The pages that follow are designed to introduce the terms, scientific concepts and broader issues involved in the situation

CARBON DIOXIDE LEVELS

Less than 1%
Percentage by which greenhouse gases rose in the 10,000 years before the Industrial Revolution

33%
Percentage by which greenhouse gases have risen since the Industrial Revolution began

650,000 years ago
Last time atmospheric carbon dioxide (CO_2) was at today's levels

3 million tons
Annual CO_2 emissions released by a typical 500-megawatt coal-fired power plant

19 lbs.
Amount of CO_2 produced by burning 1 gal. of gas in a typical car

32 billion tons
Annual worldwide human emissions of CO_2

■ Briefing

VICIOUS CYCLES

The debate over whether Earth is warming up is over. Now we're learning that climate disruptions feed off one another in accelerating spirals of destruction. Scientists fear we may be approaching the point of no return

TIME graphic by Joe Lertola; reported by Missy Adams

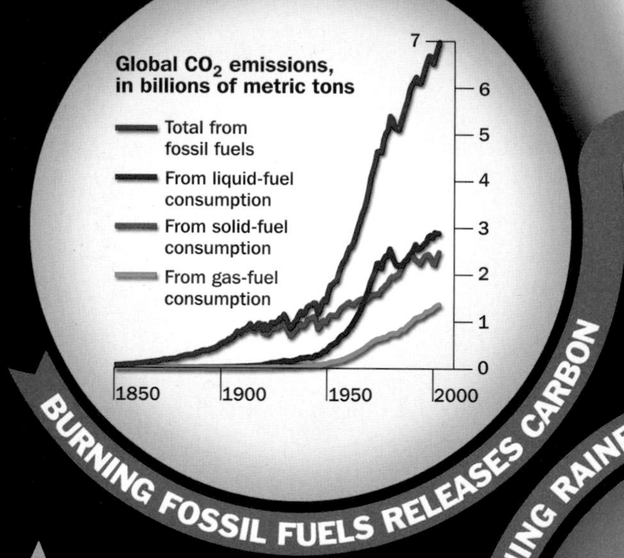

Without the **greenhouse effect,** life o Earth would not be possible. Energy fr the sun is absorbed by the planet and radiated back out as heat. Atmospher gases like **carbon dioxide** trap that he and keep it from leaking into space. That's what keeps us warm at night.

But as humans pour ever increasing amounts of greenhouse gases into the atmosphere, more of the sun's **heat g trapped,** and the planet gets a fever

SUNLIGHT HEAT approx

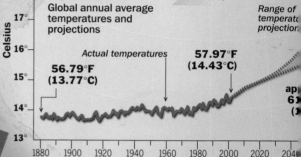

How Hot Will It Get?
Global annual average temperatures and projections

Actual temperatures

56.79°F (13.77°C)

57.97°F (14.43°C)

Range of temperatu projection

ap 6x (1

Celsius: 19° 18° 17° 16° 15° 14° 13°

1880 1900 1920 1940 1960 1980 2000 2020 2040

Global CO₂ emissions, in billions of metric tons

— Total from fossil fuels
— From liquid-fuel consumption
— From solid-fuel consumption
— From gas-fuel consumption

7 6 5 4 3 2 1 0

1850 1900 1950 2000

BURNING FOSSIL FUELS RELEASES CARBON

▲ **FUELING THE FIRE** The amount of carbon dioxide in the atmosphere is climbing fast. Most of it comes from burning fuels for energy—gasoline in cars or coal for electricity, for example. The U.S., with less than 5% of the world's population, produces one-quarter of all greenhouse gases

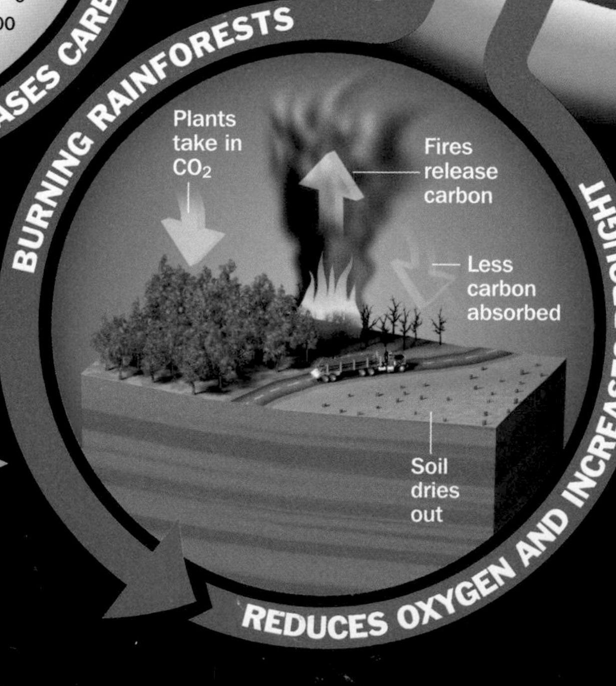

SPREADING THE PAIN Deforestation, ▶ through clear-cutting or burning, sows havoc far beyond the affected area. The fires release still more carbon into the atmosphere, fewer plants survive to convert CO₂ into oxygen, and scorched soil absorbs more heat and retains less water, increasing droughts

BURNING RAINFORESTS

Plants take in CO₂

Fires release carbon

Less carbon absorbed

Soil dries out

REDUCES OXYGEN AND INCREASES DROUGHT

MELT POLAR ICE AND PERMAFROST

Sun sea- exte
Curre 2030 (est.)

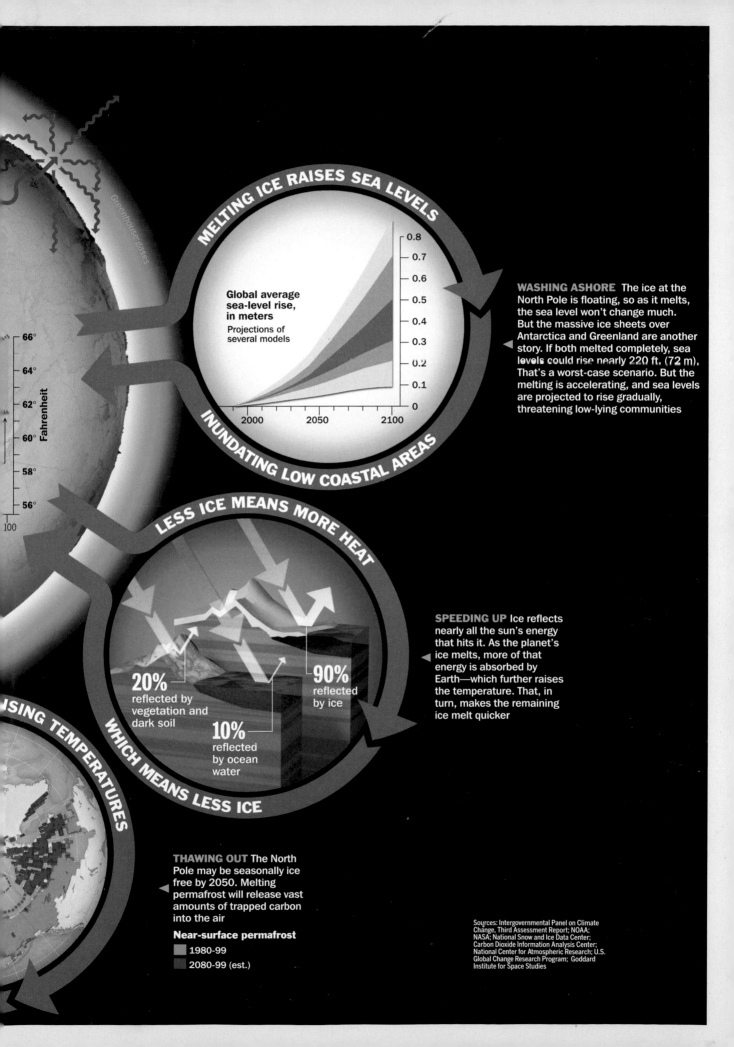

MELTING ICE RAISES SEA LEVELS

Global average sea-level rise, in meters
Projections of several models

0.8
0.7
0.6
0.5
0.4
0.3
0.2
0.1
0

2000   2050   2100

INUNDATING LOW COASTAL AREAS

WASHING ASHORE The ice at the North Pole is floating, so as it melts, the sea level won't change much. But the massive ice sheets over Antarctica and Greenland are another story. If both melted completely, sea levels could rise nearly 220 ft. (72 m). That's a worst-case scenario. But the melting is accelerating, and sea levels are projected to rise gradually, threatening low-lying communities

LESS ICE MEANS MORE HEAT

66°
64°
62°
60°
58°
56°

Fahrenheit

100

Greenhouse gases

20% reflected by vegetation and dark soil

10% reflected by ocean water

90% reflected by ice

SPEEDING UP Ice reflects nearly all the sun's energy that hits it. As the planet's ice melts, more of that energy is absorbed by Earth—which further raises the temperature. That, in turn, makes the remaining ice melt quicker

WHICH MEANS LESS ICE

RISING TEMPERATURES

THAWING OUT The North Pole may be seasonally ice free by 2050. Melting permafrost will release vast amounts of trapped carbon into the air

Near-surface permafrost

1980-99
2080-99 (est.)

Sources: Intergovernmental Panel on Climate Change, Third Assessment Report; NOAA; NASA; National Snow and Ice Data Center; Carbon Dioxide Information Analysis Center; National Center for Atmospheric Research; U.S. Global Change Research Program; Goddard Institute for Space Studies

Briefing: Frequently Asked Questions

Stand-up guy *An Adélie penguin spreads its wings in Antarctica*

What is global warming?
It is the gradual worldwide rise in surface temperature observed across the entire planet in recent decades.

What causes global warming?
The vast majority of qualified scientists believe it is caused, at least in large measure, by the greenhouse effect.

What is the greenhouse effect? How does it work?
It is the mechanism by which Earth's atmosphere behaves like the panes of glass on a greenhouse, allowing solar warmth to pass through, but preventing it from escaping back into space—thus raising the temperature inside the "greenhouse." Certain gases (especially water vapor and carbon dioxide) act like the panes of glass, allowing solar heat to pass in but not out. These are called greenhouse gases.

Is the greenhouse effect harmful?
Generally, no. It is a natural phenomenon that regulates the world's temperature and makes the planet's surface warmer than it would otherwise be. Without the greenhouse effect, life as we know it would not be possible on Earth.

So what's the problem?
The greenhouse effect is a finely tuned mechanism, and if its delicate balance is upset, Earth's surface could become dangerously hot, with potentially catastrophic results for life on the planet.

Is global warming a theory, a fact or something in between?
It is a theory, but only in the sense that gravity, relativity and evolution are theories: as in those cases, most scientists now accept the broad outlines of global warming as the most convincing explanation of an observed phenomenon, one that is strongly supported by all the available evidence. That said, there are a number of important ongoing scientific disputes about the details of the warming process.

What are the consequences of global warming?
They are impossible to predict with precision, but they could include rising ocean levels (and accompanying flooding of low-lying areas worldwide), dramatically increased storm activity, more frequent and severe droughts, the spread of deserts, the proliferation of harmful insects and the massive die-off of species whose habitats or food sources are compromised by warming.

Is human activity contributing to global warming?
Yes. Our planet's atmosphere has become overloaded with carbon dioxide (CO_2), one of the most potent greenhouse gases. Since the beginning of the Industrial Revolution, CO_2 levels in the atmosphere have spiked to levels not seen for hundreds of thousands of years. A 2007 report issued by the United Nations Intergovernmental Panel on Climate Change, the world's foremost scientific authority on the subject, declared that "warming of the climate system is unequivocal" and that there is "very high confidence" that human activity since 1750 has played a significant role in overloading the atmosphere with CO_2.

What are we doing that's contributing to the problem?
Our societies rely on power produced by burning hydrocarbons, chiefly in the form of fossil fuels such as oil, coal and natural gas. These fuels release energy when the hydrogen/carbon bond is broken, but also release large amounts of carbon dioxide and other greenhouse gases.

Has global warming occurred before?
Yes, several times in Earth's prehuman past, the atmosphere became overloaded with greenhouse gases and worldwide temperatures increased dramatically.

What do we know about earlier episodes of global warming?
Although the reasons for ancient spikes in greenhouse-gas levels and temperatures are unclear, we know with near certainty, based on geological evidence, that they occurred. We also have a reasonably clear idea of the consequences. In the Permian

Hot spot *Controversy swirls around the Alaska National Wildlife Range, where oil companies are eager to begin drilling*

Extinction, roughly 251 million years ago, the vast majority of plants and animals on the planet became extinct in the space of just 160,000 years—the blink of an eye in archaeological terms.

Does this mean global warming will trigger mass extinctions in the future?
Not necessarily, but this is at least a possibility. Many biologists believe we are now witnessing a slow-motion, planet-wide die-off of thousands of species and that global warming is one of the primary causes of these extinctions.

Do ancient examples of global warming suggest that its occurrence in our time may be natural as well?
Although this is certainly possible, it appears to be unlikely. While the causes for the elevated levels of greenhouse gases recorded in earlier instances of global warming remain unknown, scientists agree that the current rise in levels of greenhouse gases is largely of human origin.

Can global warming be stopped?
At the very least, it can be slowed, and some of its worst effects can be averted. Whether changed human behavior in the future can halt global warming entirely and even roll it back are questions that scientists are still investigating.

Is it too late to take constructive action on global warming?
No. Although our environment's natural response to stresses already registered means that some of global warming's near-term consequences will be unavoidable, these tend to be among the milder effects. There are still many opportunities, especially in the reduction of greenhouse-gas emissions, to head off the worst of global warming's predicted results.

What is the Kyoto Protocol?
It is an international agreement negotiated by 159 nations in December 1997 in Kyoto, Japan, to reduce their emissions of greenhouse gases to levels near or below their output in 1990 by the year 2012. As of mid-2007, 174 countries and the European Union, representing the vast majority of the world's populations, have agreed to attempt to meet its terms.

What is the current U.S. position

Cracking up *Drought ravages a field in India*

regarding the Kyoto Protocol?
Although Vice President Al Gore signed the Kyoto Protocol in 1997, the U.S. has never ratified the treaty or pledged to abide by its terms. U.S. critics of Kyoto view it as flawed and unfair because it calls for binding (rather than voluntary) caps on emissions, which some see as a compromise of national sovereignty. Critics also point out that it calls for some of the world's largest carbon emitters, such as the "developing nations" of China and India, to make no substantial cuts in their greenhouse-gas emissions. In turn, critics of American policy note that the U.S., with less than 5% of the world's population, is responsible for 25% of the world's hydrocarbon emissions.

How will our response to global warming affect the economy?
Attempts to adapt to global warming and mitigate its effects may require massive (and costly) adjustments in society. But many of the anticipated outcomes—cleaner air and water, more efficient fuels and greener technologies—will improve the quality of life for future generations. They will also boost economic productivity, as waves of innovation ripple through society. Some of the new, greener technologies can be expected to make fortunes for those who develop and invest in them. In any event, working to prevent the worst effects of global warming will be less expensive and

catastrophic than doing nothing and simply enduring the consequences.

Is global warming related to the "hole" in Earth's ozone layer?
They are separate but related problems. The planet's ozone layer is important because it filters out the sun's ultraviolet radiation—which is harmful to plants, animals and humans—before it reaches the surface. Ozone depletion and global warming are linked because some gases emitted by human activity, chlorofluorocarbons, for example, both trap heat (thus ratcheting up the greenhouse effect) and destroy the ozone layer. They are separate in the sense that the destruction of the ozone layer actually allows more heat to escape into space, thus partially reducing the greenhouse effect. The use of chlorofluorocarbons was banned by an international treaty in 1989.

What do we still not understand about global warming?
Many things. Scientists continue to investigate the complicated relationships that regulate climate and surface temperature on Earth. And predicting the precise timing and effects of global warming remains a work in progress. But perhaps the biggest unanswered question has more to do with politics, psychology and economics than with science: Can human beings summon the collective will to do what is necessary to meet this challenge? ∎

Briefing: Glossary

High and dry *The vast, landlocked Aral Sea was drained by Soviet-era policies that, for irrigation, diverted the rivers that fed it; here, an ancient mariner lies rusting in Uzbekistan*

air pollution The presence of chemicals and other substances in the air in amounts high enough to harm humans, animals and vegetation.

albedo The amount of heat from the sun that Earth and other celestial bodies reflect back into space. Snow and ice have high albedo, reflecting back most of the warmth they receive, while low-albedo oceans, land and plants absorb most of the heat that falls on them. As polar ice melts, the planet's overall albedo is lowered, and it soaks up more solar heat.

allocation The amount of CO_2 and other greenhouse gases that an individual user—a person, a factory, a region or nation—is permitted to discharge, based on an agreement limiting those emissions.

alternative energy Renewable sources of power not derived from burning hydrocarbons, which emit less (or no) greenhouse gas. "Old renewables" include nuclear and hydroelectric power; "new renewables" include solar and wind power.

baselines The business-as-usual level of emissions that would occur without any change in policy; useful in setting allocations and monitoring the effects of emissions-reduction programs.

biofuel A liquid or gaseous fuel made from fermented organic matter such as farm waste or crops like sugar and corn; sometimes called biogas.

carbon A chemical element essential to all forms of life on Earth, and which bonds with oxygen to form carbon dioxide, a potent greenhouse gas.

carbon cycle The worldwide passage of carbon between four "reservoirs": the oceans, the atmosphere, the ground and the bodies of all plants and animals. When one reservoir gives up too much carbon and overloads another, a delicate balance is upset, with uncertain results.

carbon dioxide (CO_2) A colorless, odorless gas produced by animal respiration, the decay of plant or animal remains and the burning of fossil fuels. Of the six principal greenhouse gases, CO_2 is the one most directly affected by human activity.

carbon footprint The amount of CO_2 created by all activities within a given sphere: a person's daily life, a city, a manufacturing process and so on.

carbon tax A charge imposed on the consumption of carbon in any form to encourage greener energy practices.

climate The long-term average, based on at least 30 years of records, of weather conditions and patterns. "Climate variability" refers to the normal range of fluctuation; "climate change" is a significant, unexpected shift in these averages.

cryosphere The frozen part of the earth's surface, including polar ice caps, continental ice sheets, mountain glaciers, lake and river ice, sea ice, snow cover and permafrost.

deforestation The conversion of woodlands to nonforest terrain; the process contributes to global warming by removing trees, which absorb CO_2 and emit oxygen. Clearing forests by setting fires pumps CO_2 into the air.

desertification The removal of plant cover, which turns fertile soil into desert. Overgrazing, deforestation, drought and extensive burning are among the causes of desertification.

El Niño A three-to-five-year cycle of warmer than usual surface temperatures in the eastern tropical Pacific Ocean, which results in disruptive changes in weather patterns.

emissions The release of substances, such as greenhouse gases, into the atmosphere. "Emissions caps" are legal limits on how much greenhouse gas a business, city or nation can emit.

energy efficiency The ratio between useful output and the energy needed to achieve it; also used to refer to reducing losses from waste in creating and consuming energy.

ethanol Another name for alcohol; a fuel source that burns more cleanly than most hydrocarbons and can be produced from organic materials like corn, sugar, soy and switchgrass.

SOURCES: ENVIRONMENTAL PROTECTION AGENCY, NASA, PEW CENTER ON GLOBAL CLIMATE CHANGE, GLOBALWARMING.ORG

evapotranspiration The loss of water from soil (by evaporation) and plants (by transpiration), increasing the amount of water vapor (a potent greenhouse gas) in the atmosphere. Evapotranspiration increases as temperatures rise.

feedback loop Any process that, in adding to global warming, further increases the rate of climate change itself. Example: higher temperatures cause greater evapotranspiration, which pumps more water vapor into the atmosphere, further increasing temperatures, which leads to more evapotranspiration and so on.

fossil fuel Combustible geological deposits (such as oil, coal and natural gas) formed over hundreds of millions of years from the decay of ancient plant and animal remains. Though a rich source of energy, they release greenhouse gases when burned.

general circulation model (GCM) A computer model of the earth's climate system that can be used to simulate human-induced climate change and predict the effects of global warming.

geothermal energy Heat transferred from the planet's molten core to water and rocks lying fairly close to the surface, where it can be tapped to produce energy.

global warming The progressive rise of the earth's surface temperature believed to be caused by the greenhouse effect and to be responsible for changes in planetary climate patterns.

greenhouse effect The process by which greenhouse gases allow incoming solar radiation to pass through the earth's atmosphere but prevent part of the outgoing heat from escaping, thus increasing temperatures.

greenhouse gas Any gas that absorbs infrared radiation from the sun as it moves back into space and traps it in the atmosphere. The six principal greenhouse gases are water vapor, carbon dioxide, methane, nitrous oxide, fluorocarbons and ozone.

hydrocarbons Organic compounds that are composed of hydrogen and carbon, including all fossil fuels.

Intergovernmental Panel on Climate Change A U.N. panel, established in 1988, that issues periodic assessments on global climate change and its effects. The IPCC is the official advisory body to the world's governments on all issues relating to climate change.

kilowatt-hour (kW-hr) A standard unit of electrical energy: 1,000 watts of current acting over a period of an hour. Often used to compare the prices of different ways of generating electricity.

Kyoto Protocol An international agreement to reduce worldwide emissions of greenhouse gases, negotiated in the Japanese city and now signed by 174 nations, but not the U.S.

natural gas Also known as methane, this fossil fuel burns much more cleanly than its hydrocarbon cousins oil and coal, but it is expensive to transport over long distances.

oxygen cycle The global exchange of oxygen (in different chemical forms) between three "reservoirs": the ground, the atmosphere and the bodies of all plants and animals.

ozone layer The layer of ozone gas in the earth's upper atmosphere that protects life on the planet's surface by filtering out harmful ultraviolet radiation from the sun.

"polluter pays" principle (PPP) The policy that countries should in some way compensate others for the effects of pollution they generate.

PPM/PPB Number of parts of a chemical found in 1 million parts (PPM) or 1 billion parts (PPB) of a particular gas, liquid or solid.

renewable energy Power obtained from sources that are essentially inexhaustible (unlike fossil fuels) and are often less polluting than nonrenewables. Examples include geothermal, wind and solar energy.

sequestration A strategy to reduce the emission of greenhouse gases like CO_2 by using either biological processes (trees) or mechanical processes, which separate carbon when generated, then store it in underground reservoirs.

sinks A natural reservoir that withdraws and traps a pollutant. Soil and plants act as carbon sinks, for instance.

urban heat island A man-made feature like a street, sidewalk, parking lot or building, which traps heat.

Dirty power *The Weisweiler power station outside Düren, Germany, burns brown coal (lignite), a form of hydrocarbon that produces far higher emissions than does black coal (anthracite)*

TROUBLE SPOTS

From depleted forests to dying reefs, distress signals dot the globe. Even in the U.S., with its relatively clean environment, excessive carbon emissions fuel global warming

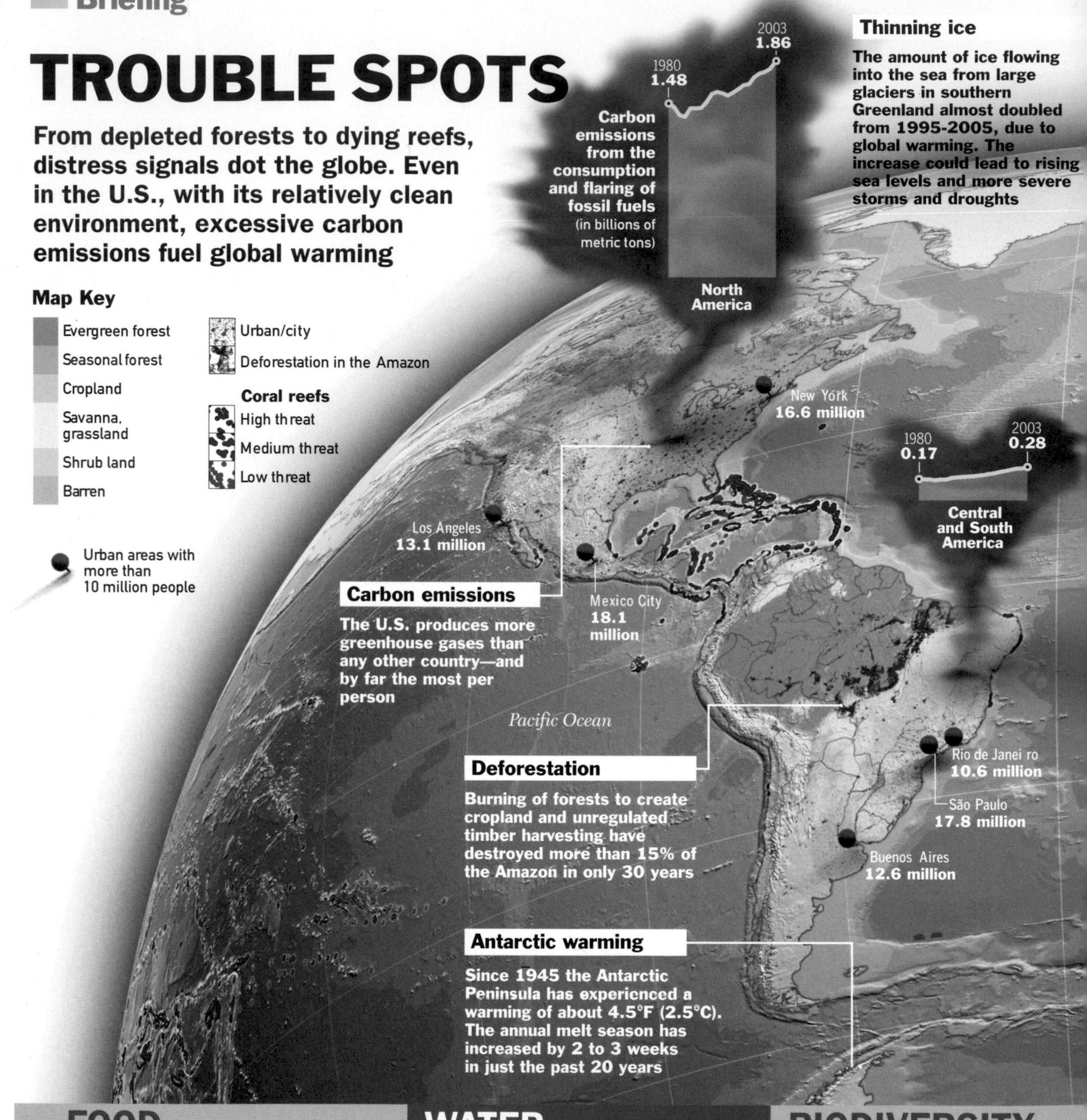

Map Key

- Evergreen forest
- Seasonal forest
- Cropland
- Savanna, grassland
- Shrub land
- Barren

- Urban/city
- Deforestation in the Amazon

Coral reefs
- High threat
- Medium threat
- Low threat

Urban areas with more than 10 million people

Carbon emissions from the consumption and flaring of fossil fuels (in billions of metric tons)

1980 **1.48**
2003 **1.86**

North America

Thinning ice

The amount of ice flowing into the sea from large glaciers in southern Greenland almost doubled from 1995-2005, due to global warming. The increase could lead to rising sea levels and more severe storms and droughts

New York **16.6 million**

1980 **0.17**
2003 **0.28**

Central and South America

Los Angeles **13.1 million**

Carbon emissions

The U.S. produces more greenhouse gases than any other country—and by far the most per person

Mexico City **18.1 million**

Pacific Ocean

Deforestation

Burning of forests to create cropland and unregulated timber harvesting have destroyed more than 15% of the Amazon in only 30 years

Rio de Janeiro **10.6 million**

São Paulo **17.8 million**

Buenos Aires **12.6 million**

Antarctic warming

Since 1945 the Antarctic Peninsula has experienced a warming of about 4.5°F (2.5°C). The annual melt season has increased by 2 to 3 weeks in just the past 20 years

FOOD

The Green Revolution helped feed developing nations in the latter half of the 20th century. But hunger continues to plague poorer countries, especially in Africa, as badly managed agriculture often leads to soil salinization and degradation

WATER

As more of the limited amount of fresh water is used each year, unequal access to supplies could produce competition and conflicts among nations. If polar ice caps continue to melt down, a major problem of the 21st century may be too much, not too little, water

BIODIVERSITY

Destruction of forests and rain forests has helped cause the worst spasm of extinctions since the dinosaurs fell vict to an asteroid impact 65 million years A 2006 report linked the extinction of species in Central America to the emis of fossil fuels

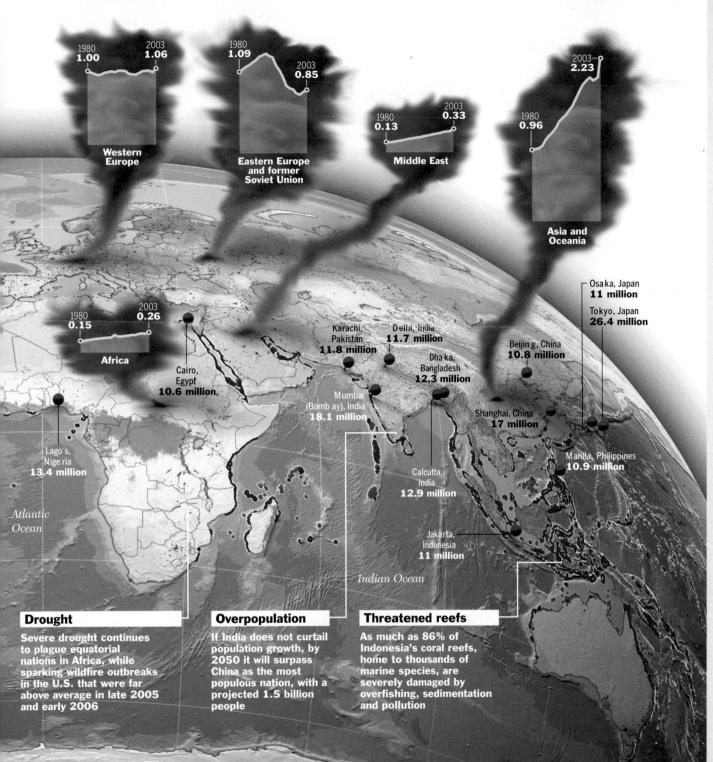

1980 1.00 **2003** 1.06
Western Europe

1980 1.09 **2003** 0.85
Eastern Europe and former Soviet Union

1980 0.13 **2003** 0.33
Middle East

2003 2.23 **1980** 0.96
Asia and Oceania

1980 0.15 **2003** 0.26
Africa

Osaka, Japan **11 million**

Tokyo, Japan **26.4 million**

Karachi, Pakistan **11.8 million**

Delhi, India **11.7 million**

Beijing, China **10.8 million**

Dhaka, Bangladesh **12.3 million**

Cairo, Egypt **10.6 million**

Mumbai (Bombay), India **18.1 million**

Shanghai, China **17 million**

Manila, Philippines **10.9 million**

Lagos, Nigeria **13.4 million**

Calcutta, India **12.9 million**

Atlantic Ocean

Jakarta, Indonesia **11 million**

Indian Ocean

Drought

Severe drought continues to plague equatorial nations in Africa, while sparking wildfire outbreaks in the U.S. that were far above average in late 2005 and early 2006

Overpopulation

If India does not curtail population growth, by 2050 it will surpass China as the most populous nation, with a projected 1.5 billion people

Threatened reefs

As much as 86% of Indonesia's coral reefs, home to thousands of marine species, are severely damaged by overfishing, sedimentation and pollution

POP./HEALTH

Life expectancy is increasing around the globe except in Africa, where AIDS and other infectious diseases have taken a toll. Lower birth rates will start to level the global population by mid-century

CLIMATE

The phaseout of chemicals called chlorofluorocarbons, achieved by a 1989 global pact, will help reduce the hole in the ozone layer. But the burning of fossil fuels will lead to hotter times in the future

ENERGY

Man's continued reliance on fossil fuels that emit carbon dioxide is extremely harmful to the planet's climate. The search for alternate fuels will be a dominant theme of 21st century science

SOURCES FOR MAP *Land use:* NASA/Boston University Department of Geography; *Urbanization:* NASA Visible Earth City Lights; U.N. Population Fund, 2000; *Amazon deforestation:* ActGlobal.org/Instituto Socioambiental; *Coral reefs:* World Resources Institute, *Reefs at Risk;* *Carbon-dioxide emissions:* Energy Information Administration; *Trouble spots:* AP; U.N. Environment Program; Global Warming Early Warning

On the Brink

Animals, plants and entire habitats are at risk from the ongoing warming of our planet. The stakes are enormous, bringing into focus the vast, intricate network that supports life on earth. From dying coral reefs to melting glaciers, from eroding permafrost to vanishing rain forests—no part of nature is an island, and the rumble of symptoms in one place can trigger a profound shock in regions thousands of miles away

Rain Forests

It's a commonplace to refer to the rain forests that bestride the globe's equatorial regions, like the one on the Caribbean isle of Dominica in the photo at left, as "the lungs of the planet." The relentless slash-and-burn tactics now being employed to harvest timber and create space for agriculture in such places as the Amazon River basin are not only sending clouds of gas into the atmosphere but also removing trees that convert carbon dioxide into oxygen, helping mitigate greenhouse-gas buildup in the atmosphere.

Rain forests might also be called "the pharmacy of the planet": they are a treasure trove of natural chemical innovation and the source of many of today's prescription drugs. Scientists fear that hundreds, even thousands, of potential miracle drugs may never be identified owing to the plundering of the rain forests.

Up in smoke *A native surveys burning rain forest in Brazil, where the isolation and vast size of the region encourage the destruction of the forest at the expense of the entire planet*

Gritty grind *Displaced persons in the Darfur region of Sudan brave a sandstorm to pump drinking water in 2004*

Deserts

Deserts play an important role in the planet's delicate balance, and we are just beginning to appreciate some of the surprising aspects of their usefulness. In recent years scientists first observed that sandstorms in north African desert regions lift minerals high into the atmosphere, where streams of fast-moving air carry them across the Atlantic Ocean to nourish the rain forests of the Amazon River basin.

As drought driven by global warming afflicts the world's hotter regions, the lives and livelihoods of millions of people are directly threatened. In many places, higher temperatures are accelerating desertification: as moisture is baked out of the soil faster, dry regions formerly at the margins of human subsistence are tipped into full-blown crisis, and once arable land becomes too dry to support agriculture.

Drought is a slow-motion, equal-opportunity killer. Many of its victims are already among the world's poorest citizens, farmers who live close to the land. Others, like the victims of "the Big Dry," a years-long drought that has afflicted thousands of farms and ranches in Australia, are the recently well-to-do descendants of pioneers who struggled to wrest harvests and wealth from a wilderness.

According to a 2005 study by the National Center for Atmospheric Research, the percentage of the earth's surface suffering drought has more than doubled since the 1970s. Among the world's hardest-hit areas are Brazil and China: more than 300,000 indigenous people have been driven from their homes as the Amazon rain forest is suffering its worst drought in more than a century. In 2007 multiple regions in China continued to suffer from the worst droughts in 50 years, which have left more than 12 million people short of drinking water and ruined the harvests of an estimated 27 million people.

Precious burden *Women carry buckets of water near Kuluku, Eritrea, in 2003, as a severe drought threatened millions with famine*

■ The Poles

The world's polar regions are the habitats most directly in the line of fire from global warming. The evidence is overwhelming and irrefutable that these long-frigid areas are not only melting down, but also that the rate of that meltdown is increasing even more rapidly than we had anticipated only a few years ago. A May 2007 NCAR study estimated that Arctic sea ice is melting more than three times faster than previously reported. In Alaska the annual mean air temperature has risen 4°F to 5°F in the past three decades, compared with an average of just under 1°F worldwide.

As a result, glaciers and ice shelves are melting, and if that melting continues, the ramifications may be felt all around the globe. If the planet's seas rise as much as 3 ft., enormous areas of densely populated land—the Nile Delta, coastal Florida, much of Louisiana, Bangladesh—would become uninhabitable. If the seas rise higher, as many scientists predict they may, major world coastal cities will be threatened, including New York City, San Francisco, Hong Kong and Tokyo. Entire climatic zones might shift dramatically: central Canada might resemble central Missouri; Alabama might be the new Antigua. Hundreds of millions of people would have to migrate from unlivable regions, amid massive social disruption.

There's more: rising seas could contaminate water supplies with salt. Warmer temperatures could widen the range of disease-carrying rodents and bugs, such as mosquitoes and ticks, increasing the incidence of insect-borne illnesses like dengue fever, malaria, encephalitis and other afflictions. For many species of wildlife, the changes could spell extinction.

Rising tides?
New York City's Manhattan Island may seem a long way from the Arctic, but polar melting is a potential threat to America's largest city

Easy does it *Scientists studying climate change aboard the Canadian coast guard icebreaker* Louis S. St.-Laurent *found in July 2006 that the ice pack was melting early, a phenomenon never noted before, making their transit unsettlingly simplified*

▪ Coral Reefs

Coral reefs are teeming undersea cities, packed with billions of tiny polyps that construct the reefs as protective exoskeletons, made from calcium. They are also nutrient factories, helping feed more than 4,000 different species of fish—not to mention sponges, crustaceans, jellyfish, mollusks, starfish, and sea urchins. And don't forget the humans: all this marine life is often the primary source of nutrition for millions of people who live near the reefs, which are located in a tropical belt around the globe that extends to roughly 30°N and 30°S of the equator. So far more than natural beauty is at risk when one-fifth of all the world's coral reefs are destroyed, as has happened since 1980. And a 2003 joint U.S.-Australian study predicts that coral reefs may disappear entirely in just a few decades.

Global warming is one of the primary culprits in this die-off. Coral reefs are such delicate biological mechanisms that a rise of just one degree in water temperature can turn these aquatic metropolises into ghost towns. In a process called coral bleaching, higher temperatures cause the polyps to eject their zooxanthellae, the algal tenants living within their tissues that provide both the breathtaking color humans admire and the nutrients the coral need to survive. The result is a reef that is deathly pale, devoid of life.

White plague *Coral in the Red Sea, top, is colorful and healthy, but many reefs around the world, like the one above in Papua New Guinea, are turning an unhealthy white*

▨ Permafrost

Permafrost is soil that has been frozen for two or more years; the ground is often, but not always, to some degree mixed with ice. Scientists estimate as much as one-fifth of the planet's land surface is covered with permafrost, which is often covered by a thin layer of soil, or "active layer," that thaws in the summer and freezes again when colder temperatures return. As with the icebound polar regions, global warming is causing unprecedented and accelerating melting of permafrost around the globe, with potentially dire results.

One of the perils of permafrost melt is shown in the picture below of the coastline of Shishmaref, Alaska, which has suffered serious erosion in recent years. When TIME reporter Margot Roosevelt visited the small Inupiaq Eskimo village that occupies a slender barrier island 625 miles north of Anchorage, she found it was "melting into the ocean." At that time, the town had lost 100 to 300 ft. of coastline, and the damage has continued. Some houses have been moved to higher ground. The ice-fishing season here used to begin in October, but late freezes of ocean ice now have delayed its start until December.

Global melting of permafrost carries a far greater danger than coastal erosion, however: the planet's permafrost regions contain enormous quantities of trapped methane, a powerful greenhouse gas, as well as other hydrocarbons. The release of large quantities of these gases would very possibly create a feedback loop that would rapidly accelerate global warming.

Edith's Checkerspot Butterfly

Researchers have documented shifts in the ranges of many butterflies due to global warming. One study looked at 35 species of nonmigratory butterflies whose ranges extended from northern Africa to northern Europe and found that two-thirds of the species had shifted their home ranges northward by 20 to 150 miles. Scientists believe many butterfly species will not survive the impact of climate change.

◼ African Elephant

Global warming not only might shrink the elephant's range within Africa but may also wreak havoc with the big animal's love life. The relative abundance—or scarcity—of food affects the social hierarchy of the herd, which in turn can determine which animals get to breed. And global warming may have a direct effect on the food supplies of the big beasts.

King Protea

The national flower of South Africa is just one of the many spectacular members of the large family of flowering plants named after Proteus, the Greek god capable of changing his shape at will. Scientists fear that more than one-third of all *Proteaceae* species could disappear by 2050.

Wild Salmon

The ritual return of salmon to their home waters to spawn is one of nature's most thrilling migrations. Now salmon populations in Alaska are at risk as melting permafrost pours mud into rivers, burying the gravel the fish need for spawning.

Prickly Pear Cactus

Across North America, warming-related changes are causing severe disruptions in the life cycles of plants. Some prickly pear cacti have lost their signature green, as seen at left, and are instead a sickly pink. Manzanita bushes, another visual icon of the American West, are currently dying back.

Harlequin Frog

A 2005 global study found that nearly one-third of the 5,743 known species of these amphibians are in danger. More than two-thirds of the 100 species of harlequin frogs in Central and South America have disappeared. Climate change seems to be making them more vulnerable to a deadly fungus.

Piñon Mouse

This tiny resident of the Southwestern U.S. lives in juniper woodlands, but in California it is heading for higher, cooler altitudes. Several other small mammals in the region have moved their homes to higher elevations over the past century.

Quiver Tree

The San people of southern Africa use the tree's hollow branches as quivers for their arrows. Now scientists have discovered that quiver trees are starting to die off in areas within their traditional range and are moving southward as equatorial temperatures keep rising.

■Adélie Penguin

Penguins, polar bears, Arctic wolves and other polar animals are the product of ages of adaptation to their frigid surroundings. As global warming melts polar ice, their habitats are endangered, and there may be no place on the planet where they can thrive. If polar melting continues at its current pace, some scientists predict polar bears will be nearly extinct by 2050. The population of Adélie penguins, above, has declined by some 33 percent in the last 30 years. The Adélie is one of two penguin species that live in Antarctica; the other is the Emperor penguin, now beloved as the star of the popular French documentary film *March of the Penguins* and the animated hit *Happy Feet*.

Are We Making Tropical Storms More Powerful?

Linking hurricanes to global warming, a growing number of scientists say that the world must brace for more intense and longer-lasting tropical storms as ocean temperatures rise

NATURE DOESN'T ALWAYS KNOW WHEN TO QUIT—AND nothing says that quite like a hurricane. Consider September 2005: the atmospheric convulsion that was Hurricane Katrina had barely finished devastating New Orleans and the Gulf Coast before its sister Rita was spinning to life out in the Atlantic. In the three weeks between them, five other named storms had lived and died in the warm Atlantic waters without making the same

headlines their ferocious sisters did. At that time, only the names Stan, Tammy, Vince and Wilma were still available on the National Hurricane Center's annual list of 21 storm names. Had the following weeks gone like the preceding ones, those names would have been used up too, and the storms that followed would have been identified simply by Greek letters. Never in the 54 years we have been naming storms has there been a Hurricane Alpha.

The year 2005 was the worst hurricane season on record in the North Atlantic, joining 2004 as one of the most violent ever. Nature took a comparative breather in 2006, at least in the Atlantic, with only nine named storms, in contrast to 28 in 2005. Five of these grew into hurricanes, but only one, Ernesto, reached the U.S., making landfall in Florida on Aug. 26 and doing relatively little damage. The Pacific Rim was less fortunate: in 2006 23 tropical storms and a record-breaking 15 typhoons (the Asian term for hurricanes) killed more than 2,500 people.

The bottom line: an undeniable trend of increasingly powerful and deadly hurricanes has now been playing out for more than 10 years. As climatologist Judy Curry, chair of the School of Earth and Atmospheric Sciences at the Georgia Institute of Technology, told TIME, "The so-called once-in-a-lifetime storm isn't even once in a season anymore." Head-snapping changes in the weather like this inevitably raise the question, Is global warming to blame? For years, environmentalists have warned that one of the first and most reliable signs of a climatological crash would be an upsurge in the most violent hurricanes, the kind that thrive in a suddenly warmer world. Scientists are quick to point out that changes in the weather and climate change are two different things. But now even skeptical scientists are starting to wonder whether something serious might be going on.

"There is no doubt that climate is changing and humans are partly responsible," Kevin Trenberth, head of climate analysis at the National Center for Atmospheric Research (NCAR) in Boulder, Colo., told TIME after Rita. "The odds have changed in

41

STRONGER SPIRALS

intensity has increased, with the number of Category 4 and 5 storms—the most powerful— nearly doubling. Given the swelling populations along the coasts, the danger from monster hurricanes like Rita and Katrina has risen dramatically

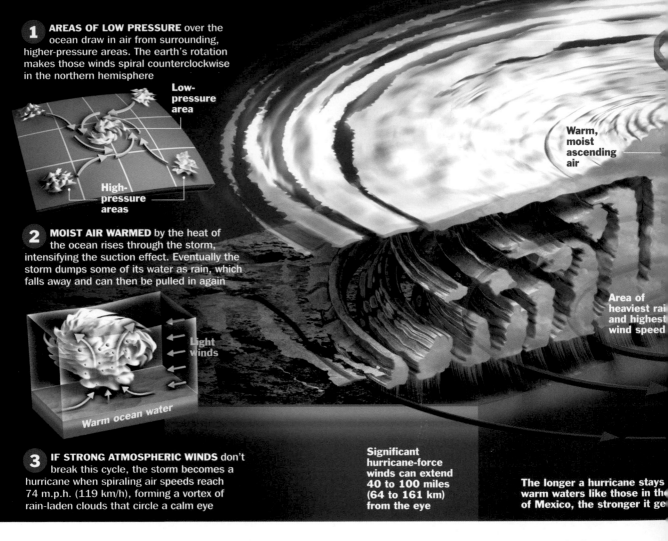

1 **AREAS OF LOW PRESSURE** over the ocean draw in air from surrounding, higher-pressure areas. The earth's rotation makes those winds spiral counterclockwise in the northern hemisphere

Low-pressure area

High-pressure areas

2 **MOIST AIR WARMED** by the heat of the ocean rises through the storm, intensifying the suction effect. Eventually the storm dumps some of its water as rain, which falls away and can then be pulled in again

Light winds

Warm ocean water

Warm, moist ascending air

Area of heaviest rain and highest wind speed

3 **IF STRONG ATMOSPHERIC WINDS** don't break this cycle, the storm becomes a hurricane when spiraling air speeds reach 74 m.p.h. (119 km/h), forming a vortex of rain-laden clouds that circle a calm eye

Significant hurricane-force winds can extend 40 to 100 miles (64 to 161 km) from the eye

The longer a hurricane stays warm waters like those in the of Mexico, the stronger it ge

---

favor of more intense storms and heavier rainfalls." Said NCAR meteorologist Greg Holland: "These are not small changes. We're talking about a very large change." In August 2006 the journal *Geophysical Research Letters* concluded that climate change offered the best explanation for the increases in sea-surface temperatures that fuel increased hurricane activity.

In 2007, Holland co-authored a study, published in the *Philosophical Transactions of the Royal Society of London,* which found that global warming's effect on wind patterns and sea temperatures has more than doubled the annual number of hurricanes in the Atlantic Ocean over the past century. "These numbers are a strong indication that climate change is a major factor in the increasing number of Atlantic hurricanes," Holland told TIME in 2007. Holland's study also noted that while 2006 was considered a quiet year, this fact is misleading: "Even a quiet year by today's standards would be considered normal

or slightly active compared to an average year in the early part of the 20th century."

Geologist Claudia Mora of the University of Tennessee at Knoxville searched a very different set of indicators to link hurricanes and global warming, studying isotopes locked in old tree rings to look for clues to past eras of heavy and light rainfall. Pair that information with global-temperature estimates for the same periods, and you can get a pretty good idea of how heat and hurricanes drive each other. "We've taken it back 100 years and didn't miss a storm," said Mora.

Searching for Every Variable

BUT DO SCIENTISTS REALLY KNOW FOR SURE? CAN MAN-MADE greenhouse gases really be blamed for the intensity of storms like Rita and Katrina? Or are there, as some other experts insist, too many additional variables to say one way or the other?

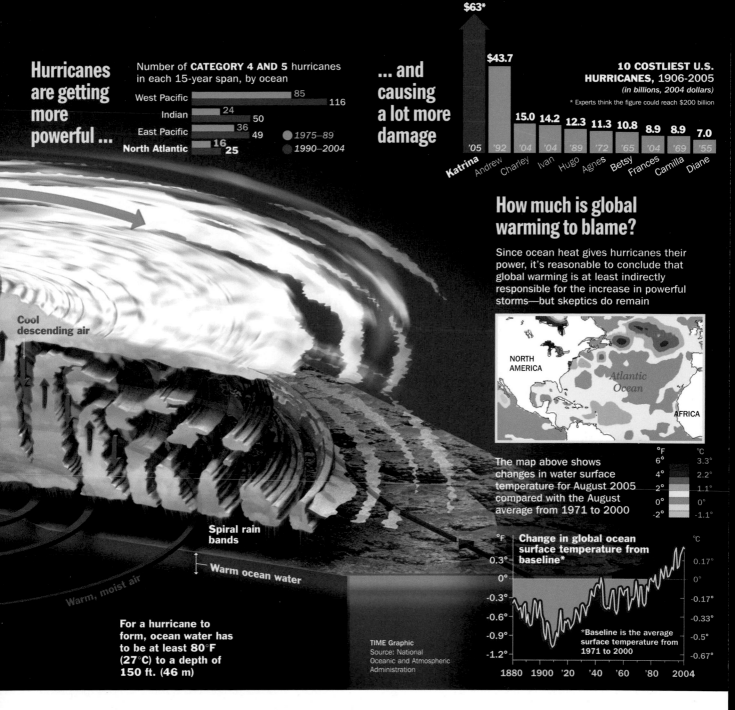

That global warming ought to, in theory, exacerbate the problem of hurricanes is an easy conclusion to reach. Few scientists doubt that carbon dioxide (CO_2) and other greenhouse gases raise the temperature of the earth's atmosphere. Warmer air can easily translate into warmer oceans—and warm oceans are the jet fuel that drives the hurricane's turbine. When Katrina hit, the Gulf of Mexico was a veritable hurricane refueling station, with water up to 5°F higher than normal. Rita too drew its killer strength from the Gulf, making its way past Southern Florida as a Category 1 storm, then exploding into a Category 5 as it moved westward.

Local events like this are not the same as global climate change, but they do appear to be part of a larger trend. Since 1970, mean ocean surface temperatures worldwide have risen about 1°F. Those numbers have moved in lockstep with global air temperatures, which have also inched up a degree. The warmest year ever recorded was 2006, followed by 2005 and 1998, with 2002, 2003 and 2004 close behind.

So that ought to mean a lot more hurricanes, right? Actually, no—which is one of the reasons it's so hard to pin down these trends. The past 10 stormy years in the North Atlantic were preceded by many very quiet ones—all occurring at the same time that global temperatures were marching upward. Worldwide, there's a sort of equilibrium. When the number of storms in the North Atlantic increases, there is usually a corresponding fall in the number of storms in, say, the North Pacific. Over the course of a year, the variations tend to cancel one another out. "Globally," atmospheric scientist Kerry Emanuel of the Massachusetts Institute of Technology told TIME, "we do not see any increase at all in the frequency of hurricanes."

But frequency is not the same as intensity, and two recent studies demonstrate that difference. In September 2005, a team

Unsafe at any speed *Courting danger, a youngster braves a bike ride as Hurricane Rita blows into Galveston, Texas, in September 2005*

of scientists that included Curry and Holland published a study in the journal *Science* that surveyed global hurricane frequency and intensity over the past 35 years. On the whole, they found, the number of Category 1, 2 and 3 storms has fallen slightly, while the number of Categories 4 and 5 storms—the most powerful ones—has climbed dramatically.

In the 1970s, there was an average of 10 Category 4 and 5 hurricanes a year worldwide. Since 1990, the annual number has nearly doubled, to 18. Overall, the big storms have grown from just 20% of the global total to 35%. "We have a sustained increase [in hurricane intensity] over 30 years all over the globe," says Holland.

Emanuel came at the same question differently but got the same results. In a study published in the journal *Nature* in 2005, he surveyed roughly 4,800 hurricanes in the North Atlantic and North Pacific over the past 56 years. While he too found no increase in the total number of hurricanes, he found that their power—measured by wind speed and duration—had jumped 50% since the mid-1970s. "The storms are getting stronger," Emanuel observed, "and they're lasting longer."

Several factors help feed the trend toward more ferocious storms. For example, when ocean temperatures rise, so does the amount of water vapor in the air. A moister atmosphere helps fuel storms by giving them more to spit out in the form of rain and by helping drive the convection that gives them their lethal spin. Warm oceans produce higher levels of vapor than cool oceans—at a rate of about 1.3% more per decade since 1988, according to one study—and nothing gets that process going better than greenhouse-heated air. "Water vapor increases the rainfall intensity," NCAR's Trenberth reported. "During Katrina, rainfall exceeded 12 inches near New Orleans."

It's not just warmer water on the surface that's powering the hurricanes; deeper warm water is too—at least in the Gulf of Mexico. Extending from the surface to a depth of 2,000 ft. or more is something scientists call the Loop Current, a U-shaped stream of warm water that flows from the Yucatán Straits to the Florida Straits and sometimes reaches as far north as the Mississippi River delta. Hurricanes that pass over the Loop typically get an energy boost, but the extra kick is brief, since they usually cross it and move on. But Rita and Katrina surfed it across the Gulf, picking up an even more powerful head of steam before slamming into the coastal states. Even if those unlucky beelines had been entirely random, the general trend toward warmer Gulf water may have made the Loop even deadlier than usual.

"We don't know the temperature within the Loop Current," said Nan Walker, director of Louisiana State University's Earth Scan Laboratory. "It's possible that below the surface, it's warmer than normal. This needs to be investigated."

Other greenhouse-related variables may also be fueling the storms. Temperature-boosting carbon dioxide, for example, does not linger in the atmosphere forever. Some of it precipitates out in rain, settling in part on the oceans and sinking at least temporarily out of sight. But the violent frothing of the water caused by a hurricane can release some of that entrained CO_2, sending it back into the sky, where it resumes its role in the warming cycle. During Hurricane Felix in 1995, measurements taken in one area the storm struck showed local CO_2 levels spiking 100-fold.

Varying Measurements, Hidden Biases

SO, ARE HURRICANES ACTUALLY SPEEDING THE EFFECTS OF GLOBal warming and thus spawning even more violent storms? That's a matter of some dispute. While many scientists agree that this outgassing process goes on, not everyone agrees that it makes much of a difference. "The amount of CO_2 given off is fairly insignificant in terms of the total CO_2 in the atmosphere," says atmospheric scientist Chris Bretherton of the University of Washington in Seattle. "I am fairly confident in saying that there is no direct feedback from hurricanes."

Thus scientific uncertainty enters the debate—a debate already intensified by the political passions that surround any discussion of global warming. The fact is, there is plenty of room for doubt on both sides of the argument. Chris Landsea, a science and operations officer at the National Hurricane Center in Miami, is one of many experts who believe that global warming may be boosting the power of hurricanes—but only a bit, perhaps 1% to 5%. "A 100-m.p.h. wind today would be a 105-m.p.h. in a century," he says. "That is pretty tiny in comparison with the swings between hurricane cycles."

Skeptics are also troubled by what they see as a not inconsiderable bias in how hurricane researchers collect their data. Since most hurricanes spend the majority of their lives at sea—some never making landfall at all—it's impossible to measure rainfall precisely and therefore difficult to measure the true intensity of a storm.

What's more, historical studies of hurricanes like Emanuel's rely on measurements taken both before and during the era of satellites. Size up your storms in radically divergent ways, and you're likely to get radically divergent results. Even after satellites came into wide use—adding a significant measure of reliability to the data collected—the quality of the machines and the meteorologists who relied on them was often uneven.

"The satellite technology available from 1970 to 1989 was not up to the job," says William Gray of Colorado State University. "And many people in non-U.S. areas were not trained well enough to determine the very fine differences between, say, the 130-m.p.h. wind speed of a Category 4 and, below that, a Category 3." (Both Gray and Landsea are critical of Holland's 2007 study and its conclusions, while Emanuel supports both.)

There's also some question as to whether there's a subtler, less scientific bias going on, one driven not by the raw power of the storms but by where they do their damage. Hurricanes that claw up empty coasts don't generate the same headlines as those that strike the places we like to live—and increasingly we like to live near the shore. The coastal population in the U.S. jumped 28% between 1980 and 2003. In Florida alone, the increase was a staggering 75%. Even the most objective scientists can be swayed when whole cities are being demolished. "If you've got a guy shooting a machine gun but he's not shooting toward your neighborhood, it doesn't bother you," meteorologist Stan Goldenberg of the National Oceanic and Atmospheric Administration in Key Biscayne, Fla., told TIME after Rita.

Even correcting for our tendency to pay more attention to what is happening in our backyard, however, the global census of storms and the general measurement of their increasing power don't lie. And what those measurements tell scientists is that this already serious problem could grow a great deal worse—and do so very fast. Americans and their leaders have ignored greenhouse warnings for years, piling up environmental debt the way we have been piling up fiscal debt. The problem is, when it comes to the atmosphere, there's no such thing as creative accounting. If we don't bring our climate ledgers back into balance, the climate will surely do it for us. ∎

Deus ex machina *Americans were stunned as Hurricane Katrina flooded New Orleans, stranding thousands. Some were rescued by helicopter*

Exhausted *No one, including the planet, profits from breathing carbon-heavy emissions from today's gasoline-powered vehicles*

When Warming Affects Health

Feeling the heat? As hotter temperatures create more severe weather patterns, health experts fear that our supplies of air and water may be tainted, while insects may spread more disease

IT'S A FAIR BET THAT GLOBAL WARMING IS GOING TO LEAD to a rise in human sickness and death. But it's a bit too soon to predict the specifics. We can be pretty sure that as average temperatures climb, there will be more frequent and longer heat waves of the sort that contributed to the death of at least 20,000 Europeans in August 2003. Other predictions are more tenuous. For example, rising temperatures could—if rainfall and other conditions are right—result in larger mosquito populations at higher elevations in the tropics, which could in turn contribute to the spread of malaria, dengue and other insect-borne infections. Early indications are not encouraging. The World Health Organization (WHO) believes that even the modest increases in average temperature that have

occurred since the 1970s have begun to take a toll. Climate change is responsible for at least 150,000 extra deaths a year—a figure that will double by 2030, according to WHO's conservative estimate. As with so many public-health issues, a disproportionate part of the burden appears to be falling on the poorest of the poor. That doesn't mean, however, that the comparatively wealthy—who account for more than their share of greenhouse-gas emissions—will escape harm. A look at three key factors affected by warming offers a hint of things to come.

Air

WE'RE USED TO THINKING OF INDUSTRIAL AND TRAFFIC POLLUtion as having a detrimental effect on air quality. But all other

46

things being equal, rising temperature by itself increases the amount of ground-level ozone, a major constituent of smog. So many studies have linked higher ozone levels to increased death rates from heart and lung ailments that many cities issue smog alerts to warn those at risk to stay indoors. You can expect more and longer alerts.

It gets worse. Higher levels of carbon dioxide favor the growth of ragweed and other pollen producers over other plants, according to Dr. Paul Epstein at Harvard's Center for Health and the Global Environment. In addition, ragweed churns out more pollen as CO_2 levels rise. Scientists have tied local spikes in asthma and allergy attacks to increases in molds and emissions from diesel engines. Apparently, the molds attach themselves to diesel particles, which deliver them more efficiently deep into the lungs. Add a plentiful helping of dust storms (from, for instance, the desertification of Mongolia or northern Africa) and a rise in drought-driven brushfires, and you have a made-to-order recipe for increasing respiratory distress worldwide.

Water

RESIDENTS OF THE U.S. GULF COAST DON'T HAVE TO BE REMINDED that water can be a killer. You can usually evacuate people ahead of a major storm, but you can't evacuate infrastructure. "Thirteen of the 20 largest cities in the world happen to be located at sea level," says Dr. Cindy Parker of the Johns Hopkins School of Public Health in Baltimore, Md. That means that where people are most at risk from floods, so are hospitals and water-treatment plants. As we have seen in New Orleans, the health effects of losing those facilities persist long after the water has receded.

Another predicted consequence of global warming is heavier downpours, leading to more floods. The immediate hazard is drowning, but the larger issue is water quality. To take just one example, more than 700 U.S. cities—most of them older communities in the Northeast, Northwest and Great Lakes area—have sewer systems that regularly overflow into water supplies during heavy rainstorms, mixing dirty and clean water and sometimes requiring mandatory boiling to make contaminated tap water safe. A heavy rainfall preceded the majority of water-borne-disease outbreaks in the U.S. over the past 60 years, says Dr. Jonathan Patz of the University of Wisconsin at Madison.

Ocean-water patterns also play a role in human health. Mercedes Pascual and her colleagues at the University of Michigan have been poring over more than a century's worth of data on cholera outbreaks in Bangladesh and tying them to detailed temperature reports of the surface waters of the Pacific Ocean. True, Bangladesh isn't anywhere near the Pacific, but the researchers are using the temperature data as an indication of a larger weather pattern called the El Niño/Southern Oscillation, or ENSO. What they have found is that the severity of an epidemic is linked to water temperature—but only in years of higher-than-normal temperatures on the ocean's surface. More alarming: as the ENSO pattern has become more pronounced since the 1970s, the association between the weather phenomenon and cholera epidemics has become even stronger.

Insects

THE NEWS HERE IS NOT ALL BAD. TICKS, FOR EXAMPLE, MAY NOT be able to survive hotter temperatures in the Southwestern U.S. And global warming is unlikely to have much of an effect on malaria, as long as you focus on lowland areas (because those regions already have so many mosquitoes). That picture may change, however, as you move upward in elevation. Malaria has seen a dramatic upswing since the 1970s in highland cities like Nairobi (around 5,500 ft. above sea level). How much of that can be tied to temperature increases—as opposed to population movement, lapses in mosquito control or the spread of drug-resistant parasites—is a matter of debate. But because each year there are at least 300 million cases accounting for more than 1 million deaths, even a small uptick in the spread or severity of malaria could be devastating.

The tricky thing about all those predictions is that you can't point to any outbreak or any individual's death and say, "This occurred because of climate change." But we do know that good public health relies on a long list of factors—the availability of doctors and nurses, effective medicines, clean water, proper sanitation—and that even today, millions of people die every year of what should be preventable diseases. With global warming, you can expect the death toll to be even higher. ∎

Health Risks

Physicians probably won't be listing "global warming" as a cause of death anytime soon. But the burgeoning signs of climate change—swelling populations of pollen-producing plants and disease-bearing insects, warmer oceans, desertification, water-supply pollution—all add up to a deadly cocktail that will challenge health experts in the future

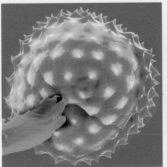

Allergies *Plants that produce substances like ragweed pollen, above, to which many of us are allergic, will flourish under the warmer conditions that are a result of global warming*

Carriers *Malaria, borne by tiny mosquitoes, kills more than 1 million people each year. Scientists fear that hot weather and heavy rainfall could make global mosquito populations explode*

Water sources *As rainfall levels increase, older cities run a risk that overflowing sewer lines may contaminate supplies of drinking water. Above, flooding lifts a manhole in the Czech Republic*

The cost of power *As China emerges from communism to become a capitalist power, its energy use is soaring. Here a coal miner in the Shanxi province takes a bag of coal back to his home*

The Burden of Asia's Giants

As China and India power up, they face a choice: become the world's worst polluters or pay the price of greener growth

YOU WOULDN'T BE READING A BOOK ABOUT GLOBAL warming if everyone lived like the average Chinese or Indian. On a per capita basis, both China and India emit far less greenhouse gas than energy-efficient Japan, environmentally scrupulous Sweden and especially the gas-guzzling U.S. (The average American is responsible for 20 times as much CO_2 emission annually as the average Indian.) There's only one problem: 2.4 billion people live in China and India, a great many of whom aspire to an

American-style energy-intensive life. And thanks to the breakneck growth of the two enormous countries' economies, they just might get there, with potentially disastrous results for the world's climate.

The statistics are mammoth and frightening. The International Energy Agency (IEA) forecasts that the increase in greenhouse-gas emissions from 2000 to 2030 from China alone will nearly equal the growth from the rest of the industrialized world. India, whose rate is growing more slowly than that of its Asian

48

rival, could see greenhouse-gas emissions rise 70% by 2025, according to the World Resources Institute. But the nearly double-digit growth rates that are responsible for those nightmare projections also present an environmental opportunity. "Anything you want to do about clean energy is easier to do from the outset," David Moskowitz, an energy consultant who has advised Chinese officials, told TIME. "Every time they add a power plant or factory, they can add one cleaner and better than before."

If China and India can muster the will and resources to leapfrog the West's energy-heavy development path, further dangerous climate change might be averted. "China and India have to demonstrate to other countries that it is possible to develop in a sustainable way," says Yang Fuqiang, vice president of the Energy Foundation in Beijing. "We can't fail."

The 1999 Kyoto accord on climate change did nothing to slow growth in China and India because, as developing countries, they are not required under the protocol to make cuts in carbon emissions—and that is not likely to change after the agreement expires in 2012. Both countries are desperate for energy to fuel the economic expansion that is pulling their citizens out of poverty, and despite bold investments in renewables, much of that energy will have to come from coal, the only traditional energy source they have in abundance.

Barbara Finamore, director of the China Clean Energy Program at the U.S.-based Natural Resources Defense Council (NRDC), estimates that China's total electricity demand will increase by 2,600 gigawatts by 2050, which is the equivalent of adding four 300-megawatt power plants every week for the next 45 years. India's energy consumption rose 208% from 1980 to 2001, even faster than China's, but nearly half the population still lacks regular access to electricity—a figure the government is working to change. "They'll do what they can, but overall emissions are likely to rise much higher than they are now," says Jonathan Sinton, China analyst for the IEA.

Not all is bleak. The NRDC is trying to help the Chinese clean up, working with their businesses to audit energy consumption and developing a fund to bankroll the installation of more efficient equipment in factories. Finamore estimates that this could eliminate the need for 3,000 new power plants over the next few decades. China also imposes higher taxes on large cars than on small ones; subsidizes wind, solar and other renewables; and has passed a law that aims to make 15% of the country's power come from renewables by 2020.

India is further behind China in developing renewable-energy sources, but the need for power is spurring innovation. India has an aggressive solar and wind industry, with one company, Suzlon, generating $1.5 billion in wind-turbine revenue in 2006. But India, with its less-developed economy, cannot as easily afford the cost of going green—or even greener. "The Indian government has not taken the problem seriously," says Steve Sawyer, a policy adviser for Greenpeace International.

Standoff: The U.S. vs. Asia

ENVIRONMENTALISM INEVITABLY TAKES A BACKSEAT TO DEVELOPMENT in China and India, but even among many green advocates there, climate change is seen as a less pressing problem than air and water pollution. There is also a widespread feeling that the developed world, which grew rich while freely spewing carbon, should take most of the responsibility for climate change. "Our issue is that, first and foremost, the U.S. needs to reduce its emissions," says Sunita Narain, director of the Center for Science and Environment in New Delhi. "It is unacceptable and immoral that the U.S. doesn't take the lead on climate change." The Bush Administration, in turn, has rejected the Kyoto agreement in part because it made developing countries exempt from emissions cuts.

The standoff between the U.S. and the Asian giants has stymied international climate-change efforts for years, but that is beginning to change—and some of the push is coming from Beijing. For most of the 2005 Montreal climate conference, the U.S. resisted any serious discussion of what should be done after Kyoto expires. But several major developing countries,

Dark days *Motorcycle riders in Inner Mongolia, China, ride through daylight obscured by smoke and smog from nearby coal-fired plants*

Cracking up *Even as China and India confront the challenge of creating cleaner energy, they are battling the symptoms of global warming. Here a woman walks on the dried-up bed of Osman Sagar, a man-made lake near Hyderabad, capital of the southern Indian state Andhra Pradesh, in 2003*

including China as a quiet but present force, supported further talks and helped break down U.S. opposition. "At the moment, China seems more interested in engaging on this issue internationally than the U.S. does," Elliot Diringer, director of international strategies for the Pew Center on Global Climate Change, told TIME in 2006.

Grow First, Clean Up Later?

BOTH CHINA AND INDIA INCREASINGLY VIEW CLIMATE-CHANGE policy as a way to address some of their immediate problems, such as energy shortages and local environmental ills, while getting the international community to help foot the bill. Thanks to poorly run plants and antiquated power grids, China and India are extremely energy inefficient. China uses three times as much energy as the U.S. to produce $1 of economic output. But that means there is a lot of room for improvement, and saving energy by cutting waste is less expensive than building new coal plants. It also reduces dependence on foreign energy and comes carbon and pollutant free. "Efficiency really is the sweet spot," says Dan Dudek, a chief economist at En-

vironmental Defense. Beijing agrees: the government aims to reduce energy intensity—the amount of energy used relative to the size of the economy—20% by 2010.

Making ambitious pledges is easy—that is what five-year plans are for—but finding the will and the funds to make them stick is trickier. One source of funding is the Clean Development Mechanism, a part of the Kyoto Protocol that allows de-

India's total energy consumption rose 208% from 1980 to 2001, even faster than China's, and nearly half the population has yet to receive regular access to electricity

Dry dock *Floating restaurants are popular with tourists in the Chongqing province in southwestern China. But a drought on the Yangtze River basin thwarted crop growth, reduced potable water supplies and left these boats high and dry in 2007*

Piped in *The village of Minqin in China's northwestern Gansu province is plagued by water shortages; its own supply is not drinkable, and desertification is taking away arable land. Water is pumped in from the nearest town, 10 km away*

veloped countries to sponsor greenhouse-gas-cutting projects in developing countries in exchange for carbon credits that can be used for meeting emissions targets. Those projects don't require any technological breakthroughs. A 2003 study by the consulting firm CRA International found that if China and India invested fully in technology already in use in the U.S., the total carbon savings by 2012 would be comparable to what

The increase in China's greenhouse-gas emissions between 2000 and 2030, scientists predict, may nearly equal the increase added by the rest of the industrialized world

could be achieved if every country under the Kyoto Protocol actually met its targets.

But that window of opportunity is closing rapidly. Every step forward that these countries take today (such as China's move to make its auto-emission regulations stricter than the U.S.'s) risks being swamped by growth tomorrow (for example, China could have 140 million cars on the road by 2020). What China and India really need to ensure green development is what the world needs: a broadly accepted post-Kyoto pact that is strict enough to make it economically worthwhile to eliminate carbon emissions. Though such cuts are off the table as of 2007, Beijing and New Delhi seem willing to discuss softer targets, such as lowering carbon intensity. But they feel that Washington must take the lead. "It is possible for these countries to achieve the growth they deserve without wrecking the climate," says Diringer. "They just can't do it on their own. It has to go through the U.S."

A modest proposal: maybe Americans can begin by living a bit more like the average Chinese or Indian—before the Chinese and Indians start living like gas-guzzling Americans. ∎

Smoke Signals

How did we get here? Even as scientists were beginning to understand the greenhouse effect, mankind began to affect it

LAME IT ON ARCHIMEDES. EVER SINCE THE GREEK mathematician experienced his "Eureka!" moment while sitting in the bath and pondering buoyancy, we've been conditioned to believe that important scientific discoveries spring from the ether, fully formed, in moments of miraculous clarity.

Global warming wasn't discovered this way. Instead, it has led scientists on a jarring pogo-stick chase, leaping across the boundaries of their disciplines and turning down numerous dead-end paths. Scholars who never dreamed they would work together—in fields as diverse as glaciology and plankton paleontology—have found themselves unlikely collaborators.

The story begins with Jean Baptiste Joseph Fourier, a polymath and titled nobleman who was nearly executed during the French Revolution. Known today for his discovery of a mathematical principle called the Fourier series, he was the first to theorize that the earth's atmosphere acts like a two-way mirror for sunlight: transparent on the way in, but reflective when the same energy (in the form of infrared radiation) tries to bounce back out into space. In 1824, Fourier coined the term greenhouse effect to describe this process.

Nobody much noticed or cared about Fourier's greenhouse until 1861, when Irish chemist John Tyndall found evidence that carbon dioxide (CO_2) and water vapor formed the inside, reflective surface of this trick mirror—and that they might be heating up the planet. "As a dam built across a river causes a local deepening of the stream," Tyndall wrote, "so our atmosphere, thrown as a barrier across the terrestrial rays, produces a local heightening of the temperature at the earth's surface."

The response of the scientific community was a collective yawn. In 1896 Swedish chemist Svante Arrhenius found that atmospheric CO_2 was rising and pegged its source to the burning of fossil fuels. "We are evaporating our coal mines into the air," he wrote. Arrhenius calculated that a doubling of CO_2 loads in the atmosphere would lead to a worldwide rise in average temperatures of several degrees. Strangely enough, he thought this a fine thing. Global warming, Arrhenius wrote, will "allow our descendants … to live under a warmer sky and in a less harsh environment than we were granted."

Or not. In the years between between Fourier and Arrhenius, the world was transformed: first by steam engines, railroads and factories, then by electricity, automobiles and ever heavier machinery. Drawing energy from moving water at its beginning, the Industrial Revolution soon outgrew hydropower and turned to hydrocarbons: initially coal, then oil and natural gas. The burning of these fuels released into the air quantities of CO_2 that were unprecedented in amount.

Later research would establish that atmospheric concentrations of carbon dioxide had held steady for at least 10,000 years before the beginning of the Industrial Revolution, which historians generally date to around 1750. But in the mere 100 years between 1850 and 1950, those levels increased more than 33%.

In 1938 Stewart Callendar, an amateur British weatherman, proposed to the Royal Meteorological Society that a widely noticed run of warm years through the 1930s might be connected to Fourier's greenhouse hypothesis. This is perhaps the first time that global warming attracted public rather than scientific notice, but both scholars and laymen shrugged off the theory.

The next piece of the puzzle arrived courtesy of the cold war. U.S. Navy nuclear submarines on patrol beneath the ice caps around the North Pole needed places to break through the ice, so their crews took detailed measurements of its thickness. Decades of these data (made public much later) revealed an unmistakable trend: the ice caps were getting thinner.

In 1957 Roger Revelle, a scientist at the Scripps Institution of Oceanography, published a paper showing that the ability of the world's oceans to absorb man-made CO_2 was limited, concluding, "human beings are now carrying out a large-scale geophysical experiment." In 1958 he obtained funding for chemist Charles Keeling to take measurements of atmospheric CO_2 levels at Hawaii's Mauna Loa volcano. Within a few years, this data showed the levels were rapidly rising.

Through the 1960s and '70s, scientists became increasingly aware of, and alarmed by, the mounting evidence of a trend toward planetary warming. The 1972 U.N. Conference on the Human Environment in Stockholm put climate change on its agenda, and 1979 marked the first World Climate Conference. Even so, the public and national leaders remained largely unconcerned. Around this time, however, the insurance industry noticed that it was paying out more money for weather-related damages. By the early 1980s, climate change had begun to attract the attention of the environmental community and mainstream media. And in 1988, the U.N. formed the Intergovernmental Panel on Climate Change (IPCC). Two years later, that group issued is first assessment: man-made emissions of greenhouse gases were likely to cause rapid, possibly catastrophic climate change in the foreseeable future.

Archimedes is said to have died in 212 B.C., after telling an invading Roman soldier to leave him alone while he worked on a problem. It remains to be seen whether human beings, having completed the long game of connect the dots that leads from Fourier's day to the present, will also suffer because they too are unwilling to be disturbed. ∎

Good News, Bad News

The Industrial Revolution was a great leap forward for mankind, as we began to harness the energy in fossil fuels to power a host of new machines and generate electricity. But the waste matter created in the process helps trap heat in the planet's atmosphere.

Something in the air
Clouds of hydrocarbon emissions billow from a Pittsburgh, Pa., steel plant in the 1890s

1859 *Edwin Drake strikes oil in Pennsylvania after drilling the world's first petroleum well. Oil, coal and other hydrocarbons soon power many of the world's big machines.*

1908 *Oil comes into its own when Henry Ford unveils the affordable Model T. Today, a typical car pumps roughly 19 lbs. of carbon dioxide into the atmosphere per gallon of gas.*

1962 *The modern environmental movement is born, with the publication of Rachel Carson's Silent Spring. Although the book dealt largely with the harmful effects of pesticides (and led to a ban on some*

especially toxic chemicals), it made people aware of the notion that human activity can harm the environment.

1970 *Environmentalism goes mainstream on April 22, with the first observance of Earth Day. This marks a crucial watershed: the environmental movement grows in size, and its focus widens from local and regional concerns to embrace a truly global agenda.*

53

Viewfinder

Three gifted photographers turn their lenses on the causes and consequences of global warming and share their stories of covering the crisis

PHOTOGRAPHS BY J. HENRY FAIR—REDUX

A grove of trees clings to its last moments of life at a coal mine near Kayford, W. Va. "What you can't see in a photograph," says Fair, "is the rapidity of the destruction. This mountaintop barely lasted a day. To get a sense of the scale of these huge machines, take a look at the full-size red pickup truck in the background."

■ **J.Henry Fair**

Like many professional photographers, J. Henry Fair shoots a variety of subjects, but in recent years he's devoted many hours to documenting the causes and symptoms of global warming, especially as reflected in the industrial practices involved in producing hydrocarbon-based fuels. His work, he says, is pulled in two directions, between art and journalism. And we'd add a third center of gravity to that list: advocacy.

On the subject of the environment, Fair's documentation of climate change and what he calls "industrial scars" have turned him into an outspoken proponent of greener lifestyles. "The trick," he says, "is getting people to look at familiar subjects with fresh eyes and associate the consequences of the dollars we spend with the damage our consumption is causing. If we can do that, everyone will go green at the flip of a switch. In these pictures, I'm trying to get us and our kids to feel the very real effects of our everyday actions, as simple as leaving a computer on or buying a roll of toilet paper. How do we teach ourselves to be conscious of those choices and work that knowledge into choosing the products we buy? This issue is all about the children: What kind of planet do we leave them?"

Slurry from a coal-mining operation flows into a holding reservoir, resulting in a photograph that bears a haunting resemblance to a nerve scan.

The blue splash of color is the process of hydroseeding, as a mixture of seeds and liquid fertilizer is sprayed where trees have been clear-cut, in a process that satisfies minimal federal regulations for mitigating strip mining. Says Fair: "In one day, an old-growth forest has been turned into a parking lot with a turquoise paint job."

Many of Fair's pictures have the seductive palette and formal rigor of an abstract expressionist painting, drawing us in with their energy and oddity, only to surprise us as their actual subject matter swims into comprehension. There are often such razor blades in Fair's eye candy, for these are images of the methodical ravaging of a planet. Many photographers provide only the most essential data in their photo captions: location, date, subject. The captions Fair provides to editors and gallery curators, in contrast, often bristle with statistics, documenting the precise price to the environment of the industrial processes he is recording.

Fair snapped this picture of a coal-burning power plant in Louisiana on a December morning in 2000, capturing an indelible portrait of hydrocarbon emissions streaming into the atmosphere and turning what might otherwise have been a clear winter vista into a study of the texture of smog. Fair's caption drives home the point, as he quotes a telling statistic from the Union of Concerned Scientists: in an average year a typical coal-burning power plant generates 3.7 million tons of carbon dioxide emissions, which are not only a primary driver of global warming and acid rain, but also contribute to the misery of those who suffer from asthma and bronchitis.

■ Portfolio

■ Gary Braasch

For most of the first years of the 21st century, photographer Gary Braasch, 63, has been on the road, traveling the world to document global warming. The result of his odyssey, *Earth Under Fire: How Global Warming Is Changing the World,* was published in the fall of 2007 by the University of California Press. In its almost 300 pages, the volume paints a wide-ranging but also finely detailed portrait of a planet in crisis. It is also a labor of love; though it includes pictures from stories assigned by magazines, Braasch paid for much of the expenses involved in his long photographic crusade. All the photos on these pages appear in *Earth Under Fire.*

Braasch earned a master's degree in journalism from McGill University, but his goal was to be a writer, not a photographer. It was only after editors complimented the photos he took to accompany his stories that Braasch decided to devote himself to the camera. Born in Nebraska, he settled in Oregon, where he recalls being stirred by the sight of old-growth forests being cut down for timber.

"Initially, I was interested in taking 'beauty shots' of nature," Braasch says. "But the eruption of Mount St. Helen's in 1980 was a turning point for me. For the first time, I saw a nature story dominating the headlines and profoundly affecting human events. Since that time, I have devoted most of my time to covering the environment as it affects our lives."

Passionate on the subject of global warming, Braasch keeps a close eye on the scientific journals that publish new findings in the area, then approaches scientists to ask permission to join them in the field. The result, as seen in *Earth Under Fire* and in these pages, is striking visual documentation of a world dealing with a crisis that knows no borders.

Zoologist Steve Williams and a research associate survey the rain forest at the Wet Tropics World Heritage Area in Queensland, Australia, where they found that climate change was forcing animals to more elevated habitats.

Braasch took the two pictures above of Mt. Hood in Oregon in 1986 and 2002; it is one of a number of before-and-after diptychs in Earth Under Fire. The loss of glacial ice near the crest of the mountain over this relatively brief time span (in geological terms) is obvious: scientists estimate Mt. Hood has lost some 40% of its glacial ice since 1900.

It was only after he had taken a number of other such time-lapse photos, duplicating the scenes shot by other photographers, that Braasch realized he could draw on his own 1986 picture of Mt. Hood to show the changing conditions on the peak. "It's quite an adventure," he relates, "to traipse around mountains, rain forests and glaciers trying to find the exact vantage point where another photographer stood 20 or 50 years before." Braasch finds such "before" shots by combing through racks of old postcards in local markets and poring over old scientific journals.

"I am not an underwater photographer," Braasch declares. As a professional, he has high respect for his colleagues who shoot undersea life, a task that is demanding on several levels at once, requiring special equipment, adequate funding, scuba expertise, good lighting, calm seas—and a strong dose of luck in getting aquatic life to "pose" for a picture. "It's an amazing skill," says the photographer. "Those guys are lucky to get one good shot a day."

Yet Braasch's photographic survey of climate change demanded that he document the threat to the world's coral reefs posed by higher ocean temperatures, so he trained to earn his qualification as a scuba diver, then visited the world's most extensive coral habitat, Australia's Great Barrier Reef, to record its condition. His one regret: "I wasn't able to see and document the bleaching that is the telltale sign of a dying reef."

"The picture above may seem to be an indictment of pollution," says Braasch. "But the pipes in the photo are simply storm drains that divert rainfall into the ocean. The real reason I went to the Delaware Bay shoreline was to shoot the migrating birds you see here along the waterline." The beach is on the main route of the Atlantic flyway, one of the world's great avian migratory highways. Like many Atlantic beaches in the U.S., it is now eroding as sea levels rise—and that means there is less space for shore birds to hunt the horseshoe crabs that make up the bulk of their diet. Rising seawater levels pose a danger to many more plants and animals along the shores of the U.S. East Coast, threatening to inundate freshwater estuaries with salty ocean water.

In Bangladesh, top right, Braasch found people for whom climate change is a current crisis rather than a theoretical risk. The people in the village of Charkamli, he says, were prepared for riverbank erosion but not for the rate at which it is now progressing: villages have been abandoned, and an entire mosque was lost when rising waters changed the course of a river.

On the South Pacific archipelago of Tuvalu, right, Braasch recorded kids waiting out a tidal flood. A plan for evacuating the island nation is now in place, as rising ocean levels threaten its supply of freshwater. "A sovereign nation is now becoming uninhabitable through natural action," says Braasch. "That's a new idea for the human race to contemplate."

■ **Peter Essick**

He may not have been born with a camera in his hand, but Peter Essick, 49, was raised by a father who taught science and loved the outdoors. An amateur shutter-bug since his high school days, Essick received a master's degree in photo-journalism from the University of Missouri Journalism School, then worked at several newspapers before coming to rest where he seemed preordained to be, at *National Geographic* magazine. It was a 1993 assignment to shoot pictures for a magazine-length story on water that first brought Essick into the working orbit of scientists who were tackling global environmental issues.

Many of the pictures shown on these pages were taken while Essick was covering climate issues for *National Geographic*. "When I first started taking these pictures," he says, "I knew a bit about global warming, but not much. After talking to the scientists and seeing the evidence add up—in lakes, along coast-lines, in mountains and on coral reefs—I began to grasp the extent of the problem. It was really the scientists themselves and their urgency that convinced me. When I first started covering this subject, the debate was still in the 'Is it for real?' area. Now people have advanced beyond that. They want to address the issue."

A scientist with the Global Observation Research Initiative in Alpine Environ-ments (yes, the acronym is GLORIA*) lays out a 1-meter quadrat , a measured rectangle used in field biology to register the population of flora in a given area. Ten years before, scientists took measurements of plant life here on Mount Schrankogel, Austria; the new findings showed that grasses and native wild-flowers were moving steadily upslope and that some flowers that grow only near the mountain's summit are threatened with extinction.*

Research engineering associate Jason Sun monitors displays in the Center for Computational Visualization at the University of Texas, Austin. The center's sophisticated supercomputer uses data on ocean circulation and temperatures provided by the Scripps Oceanographic Institute to create a visual simulation of real-time ocean conditions around the world, as well as to perform simulations of the effects of climate change over time.

PHOTOGRAPHS BY PETER ESSICK—AURORA

Brrrrr! Climatologist Lonnie Thompson holds an ice core from the Quelccaya Ice Cap in Peru in a cold-storage freezer at Ohio State University. Thompson and his colleagues keep cores from 20 years of drilling in such high-altitude tropical glacier fields at −20°C for analysis. It was Thompson, Essick says, who convinced the photographer of the urgency of the global-warming crisis over the course of a weeklong expedition to Peru to collect ice-core samples. By collecting cores where the ice is very deep, as in polar regions, scientists are able to retrieve weather data dating back tens of thousands of years and thus trace the planet's changing climate conditions.

Paleoclimatologist Jeff Dorale, left, of the University of Iowa holds a stalagmite from Crevice Cave near Perryville, Mo. Like ice-core samples, stalagmites contain records of the past. Dorale collects them, cuts them in half in the lab and dates their layers with radioactive isotopes. Carbon levels within the stalagmite help Dorale determine the amount of vegetation in the ground over the cave at different time periods; oxygen isotopes reveal ancient temperatures. The stalagmite is 130,000 years old.

Christian Both, top right, conducts research on the pied flycatcher in Hoge Veluwe National Park, the Netherlands. The birds migrate from Africa each year; Both found that their arrival date, believed to be based on the position of the sun, has not changed, but warmer spring temperatures are making the birds' chief food, winter moth caterpillars, emerge earlier. In response, the flycatchers are now laying their eggs more quickly after their arrival.

Essick photographed biologist Bill Fraser, right, hanging out with a few friends on Torgerson Island, Antarctica. Fraser is capturing Adélie penguins to put satellite tags on them. When the scientist first began monitoring the penguin flock some 30 year ago, there were 1,000 breeding pairs on the island; now there are fewer than 20. The sea ice that supports the krill on which the penguins feed is vanishing.

Plant ecologist Tad Day, left, gets up close and personal with something new under the sun: plant life near Palmer Station, Antarctica. In 1995 scientists reported there were no plants here at all, since the ground was completely covered by ice pack. By 1999 researchers found 23 cushion plants and 94 grass plants, and by 2004 there were 294 cushion plants and 5,129 grass plants.

Deep defense *The Metropolitan Area Outer Underground Discharge Channel is a four-mile underground conduit in northwest Tokyo that was built to drain floodwaters, as the island nation prepares for much wetter years ahead*

Solutions

JOE NISHIZAWA

Big problems demand big answers, so governments, scientists, power companies and concerned citizens are tackling the problem of climate change on a variety of fronts, searching for ways to reduce pollution, generate cleaner energy and adapt to the challenges posed by an ever warmer planet

To Catch the Wind

Generating electricity from the power of the wind is one of the few forms of alternative energy that make economic sense at present: when a federal tax subsidy is factored in, wind power is competitive with the least expensive—and most polluting—forms of traditional fuel, like coal. And it doesn't produce a wisp of carbon exhaust. The power industry is paying heed: by 2004, U.S. wind farms like this behemoth outside Palm Springs, Calif., were creating twice as much of the U.S. power supply as they did in 2000, an amount equivalent to the burning of 9 million tons of coal.

So why isn't wind power catching on even more? Apart from the iffy nature of relying on the breeze for steady power, some folks bemoan the birds and bats that are Cuisnarted by giant wind turbines, while others who live near wind farms grumble about noise, unpleasant views and declining property values. But as global warming heats up— well, you don't need a weatherman to know which way the wind blows.

SOLUTIONS

Giant **wind farms** that harness the power of moving air are

roducing more and more of the world's power supply

Plugging into the Power of the Planet

Solar and wind power rely upon factors that can't be predicted: ample sunlight, a sustained breeze. But geothermal power, which taps the scorching temperatures that radiate outward from the earth's core, is theoretically available everywhere on the planet—the only variables are how deep you need to dig before finding it, and at what cost. And as prices for gasoline and natural gas creep ever upward, geothermal seems increasingly like a bargain. A 2006 report by the Massachusetts Institute of Technology indicated that with a $1 billion investment in the coming years, geothermal energy could supply 10% of all U.S. power needs by 2050, on a par with nuclear and hydroelectric power. The world's role model in this process will be Iceland, whose Nesjavellir geothermal plant is shown here; 90% of the nation's power is generated from geothermal and hydroelectric sources.

SOLUTIONS

In Iceland, **geothermal power,** which taps heat generated a

RAGNAR SIGURDSSON—GETTY IMAGES

he fiery core of the planet, is a major source of energy

Guarding a Scepter'd Isle

Rising water levels are one of the most potent threats posed by global warming. In response Britain has doubled its spending on flood prevention. The most visible evidence of the danger is the Thames Barrier, a set of hulking yet beautiful floodgates that bisect the waterway 11 miles downriver from central London. When the barrier became operational in 1983, 30 years after the massive flood that spurred its construction, planners expected that it might have to close once or twice a year to keep ocean storm surges from inundating the capital. In the past decade, however, extreme weather has caused the barrier to be closed an average of 10 times each year.

SOLUTIONS

The Thames Barrier floodgates protect London from high

vaters. They are used far more often than planners expected

Let There Be Energy—from Light

The amount of solar energy that bathes the earth in a single hour could, if harnessed, supply energy needs around the world for a year. The technology to harvest even a tiny fraction of that energy at a reasonable cost does not yet exist—but we're getting closer. It now costs seven times as much to create electricity from solar panels as it does from burning coal. Why the high price tag? Silicon is expensive, and it's needed to make the photovoltaic cells that convert sunlight to energy, as seen here near San Luis Obispo, Calif. Another factor: new technology and the infrastructure to support it require substantial up-front investments. Even so, the price of solar energy has declined steeply in the past few years and will continue to fall. Add tax breaks and other forms of official support, along with economies of scale as the business matures, and solar power, which produces no carbon emissions or other greenhouse gases, begins to look competitive with cheap-but-dirty hydrocarbons.

SOLUTIONS

Reversing the polarity of the climate crisis, solar power

aws energy from the warming sunlight that threatens us

Adapting to A Warmer Planet

Changing climate, changing lives. As we focus on reducing the emissions that are a main cause of global warming, we must also begin to shape our lives to adjust to a warmer, wetter planet

PERHAPS IT WAS ALWAYS TOO OPTIMISTIC TO BElieve that human beings would be responsible stewards of the planet. We may be the smartest of all the animals, endowed with exponentially greater powers of insight and abstraction, but we're animals all the same. That means we can also be shortsighted and brutish, hungry for food, resources, land—and heedless of the mess we leave behind trying to get them.

And make a mess we have. If droughts and wildfires, floods and crop failures, collapsing climate-sensitive species and the images of drowning polar bears didn't quiet most of the remaining global-warming doubters, the hurricane-driven destruction of New Orleans did. Dismissing a scientist's temperature chart is one thing. Dismissing the death of a major American city is something else entirely. What's more, the temperature is continuing to rise. "The science," says Christine

different results. You can choose a hybrid vehicle, but simply tuning up your car and properly inflating the tires will help too. Buying carbon offsets can reduce the impact of your cross-continental travel, provided you can ensure where your money's really going. Planting trees is great, but in some parts of the world, the light-absorbing color of leaves causes them to retain heat and paradoxically increase warming.

Even the most effective individual action, however, is not enough. Cleaning up the wreckage left by our 250-year industrial bacchanal will require fundamental changes in a society hooked on its fossil fuels. Beyond the grass-roots action, larger tectonic plates are shifting. Science is attacking the problem more aggressively than ever. So is industry. So are architects and lawmakers and urban planners. The world is awakened to the problem in a way it never has been before. Says Carol Browner, onetime Administrator of the EPA: "It's a sea change from where we were on this issue."

For years, global warming was discussed in the hypothetical, as a threat mankind might face in the distant future. Now it is increasingly regarded as a clear, observable fact. This sudden shift means that all of us must start thinking about the many ways global warming will affect us, our loved ones, our property and our economic prospects. We must think— and then adapt accordingly.

Adaptation and Mitigation

WHEN CLIMATE SCIENTISTS USE THE WORD *ADAPTATION*, THEY are referring to actions intended to safeguard a person, community, business or country against the effects of climate change. Its complement is *mitigation*—any measure that will reduce greenhouse-gas emissions, such as drawing power from a wind turbine rather than a coal-fired power plant. Mitigation addresses the front end of the global-warming problem; by cutting emissions, it aims to slow rising temperatures. Adaptation is the back end of the problem—trying to live with the changes in the environment, society and the economy that global warming will continue to generate.

For years, adaptation was overlooked or disparaged in policy circles; many complained that even discussing it was a sellout that gave governments and others an excuse not to act. Today adaptation has become an accepted part of the discussion. A report from the Intergovernmental Panel on Climate Change (IPCC), released in April 2007 in Brussels, makes it official. "Adaptation to climate change is now inevitable," says Roger Jones of the Commonwealth Scientific and Industrial Research Organization in Australia, a co-author of the IPCC report. "The only question is whether it will be by plan or by chaos." Jones, like the other contributors to the IPCC report whom TIME interviewed, spoke only for himself, not his organization.

The need for adaptation is rooted in the unhappy fact that we can't turn global warming off, at least not anytime soon. The momentum of the climate system—carbon dioxide remains in the atmosphere for decades, while oceans store heat for centuries—ensures that no matter how much humanity cuts greenhouse-gas emissions, our previous emissions will keep warming the planet for decades. Even if we were to magically stop all emissions today, "temperatures will keep rising, and all the impacts will keep changing for about 25 years," says Sir David King, chief science adviser to the British government.

Todd Whitman, former Administrator of the Environmental Protection Agency (EPA), "now is getting to the point where it's pretty hard to deny." Indeed it is. Atmospheric levels of carbon dioxide (CO_2) were 379 parts per million (p.p.m.) in 2005, higher than at any time in the past 650,000 years. Of the 12 warmest years on record, 11 occurred between 1995 and 2006.

So if the diagnosis is in, what's the cure? A crisis of this magnitude clearly calls for action that is both bottom-up and top-down. Though there is some debate about how much difference individuals can make, there is little question that the most powerful players—government and industry—have to take the lead.

Individuals can indeed move the carbon needle, but how much and how fast? Different green strategies, after all, yield

Sea change *Fishing for prawns in the salty waters of Sunderbans, in southern Bangladesh, where ocean storm surges are flooding coastlines*

So while we strive to make our economies greener, we must also mount a major new effort to strengthen our resilience against the impact on the climate that our past carbon emissions have set in motion.

Public discussion of global warming in the U.S. is years behind the rest of the world, and adaptation is no exception. "You can't adapt to a problem you don't admit exists," notes Richard Klein of the Stockholm Environment Institute, another IPCC co-author. The U.S. has only recently acknowledged global warming, while other countries are already taking concrete action to prepare for its impact. The Netherlands has some of the strongest flood defenses in the world and is making them stronger. Britain has doubled spending on flood and coastal-defense management, to about $1 billion a year. France, Spain and Finland have launched less ambitious adaptation initiatives. Even Bangladesh, one of the world's poorest nations, is taking action.

Nevertheless, adaptation has implicitly emerged on the American agenda, thanks to Hurricane Katrina. The earth's weather system is too complex to pin blame for Katrina definitively on global warming. But unusually strong hurricanes like Katrina are exactly what scientists expect to see—along with fiercer heat waves, harsher droughts, heavier rainfalls and rising sea levels—as global warming intensifies. If the nation is serious about reviving New Orleans and its neighbors, it must make them as resilient to global warming as possible. "We have to fight for New Orleans," says Beverly Wright, director of the Deep South Center for Environmental Justice at Dillard University in New Orleans. (Her house took on 8 ft. of water after Katrina.) "If we're vigilant, we can make New Orleans the safest coastal city in the world and then use it as a model for

how the rest of the country can get ready for global warming."

Unfortunately, New Orleans today remains far from that ideal. Robert Bea, a professor at the University of California, Berkeley, and former oil-industry engineer, co-authored a landmark report for the National Science Foundation that analyzed why the Federal Government did such a poor job of protecting Louisiana before and after the storm. Most of the problems he identified persist, he says. And that is not Louisiana's problem alone, Bea emphasizes. The Army Corps of Engineers announced early in 2007 that 122 major levee systems are less than safe; those levees will face greater stresses with global warming. Extra-strong hurricanes will threaten cities along the entire Gulf and Atlantic coasts. Scientists say New York City is long

Gatekeeper *The Thames Barrier at work, controlling the flow of the river during an ocean surge. The gates will be raised another 12 in.*

82

The Netherlands: instead of trying to contain floods that threaten the land, much of which is below sea level, the Dutch are planning to allow predesignated areas to flood

overdue for a major hurricane; global warming raises the odds.

"All Americans should look carefully at what is and isn't happening in New Orleans," says Mark Davis, professor of environmental law at Tulane University in New Orleans. "If we can't marshal the money, technology and political will to succeed here, I wouldn't be confident we'll do much better in your part of the country either." Meanwhile, Americans can look abroad for examples of how to prepare for climate change.

The Netherlands

IT'S NO SURPRISE THAT THE NETHERLANDS HAS ONE OF THE BEST records in the world on adaptation. The Dutch have been coping with their low-lying location for nearly 800 years. Dutch law requires that river defenses deliver so-called 1-in-1,250 protection—that is, limiting the odds of catastrophic system failure and consequent flooding to 1 in 1,250 years. (By comparison, New Orleans' defenses offered 1-in-100-years protection.)

To maintain this level in the face of greater anticipated flows down the Rhine River (thanks in part to accelerated snowmelt in the Alps), the Dutch are radically revising traditional flood-management thinking. Instead of trying to contain floods, they will accommodate the extra water flow by allowing predesignated areas to flood. The strategy is called Living with Water. Near Nijmegen, the oldest town in Holland, a sparsely populated strip of land that is home to farms and a nature reserve will be allowed to flood to spare the more heav-

ily populated areas downstream. Birds in the nature preserve can fly away until the waters recede, but not homeowners, who have protested. One lesson, says Bas Jonkman, an adviser to the Dutch Ministry of Water Management, is that "society must recognize that there will be losers from adaptation, and they must be compensated."

The greatest flood danger to the Netherlands comes from the North Sea, which is more powerful and unpredictable than Dutch rivers. So Dutch law has historically required North Sea defenses to deliver a 1-in-10,000-years level of protection. "And now the Parliament wants to raise the North Sea standard to a 1-in-100,000-years level of protection," says Pier Vellinga, a senior government adviser and professor at Wageningen University and Research Center. Vellinga calculates that to maintain this higher level of protection, the Netherlands would have to commit about 0.2% of its GDP annually—some $1.3 billion.

The Dutch are straightforward about making adaptation to global warming a high priority. The alternative is the prospect of losing their coastal cities altogether. ("We Are Here to Stay" is the accompanying public slogan.) "We want foreign visitors and investment to keep coming to the Netherlands," Vellinga says, "so we must assure them this will remain a safe place."

Britain

THE MOST VISIBLE EXAMPLE OF THE BRITISH COMMITMENT TO adaptation is the Thames Barrier, a set of hulking but beautiful

Dutch dams *The Oosterscheldekering is part of the Delta Works barrier. The Netherlands is planning now to handle bigger storms to come*

silver floodgates that stretch across the namesake waterway about 11 miles downriver from central London. When the barrier became operational in 1983—30 years after a massive flood, driven by a storm surge, killed 58 people and motivated its construction—planners expected that it might have to close once or twice a year to keep ocean storm surges from inundating London. In the past decade, however, the barrier has been closing an average of 10 times a year.

"The barrier was initially designed to offer a 1-in-2,000-years level of protection," says Chris West, an Oxford University professor, trained as a zoologist, who is now directing the United Kingdom's Climate Impacts Program. "But sea-level rise is projected to reduce that to a 1-in-1,000-years level by 2030." In response, the British government is prepared to add 12 in. of protection on top of the existing floodgates—a contingency built into its original design—and to keep building patches and extending the barrier as necessary. Planners in Britain assume the system will have to be replaced within 100 years, but they don't yet know with what.

Adaptation isn't just about building a stronger physical infrastructure. A new urban village is being planned 120 miles north of London that will bring together mitigation and adaptation. "Bilston Village will not only be a low-carbon-energy

Bangladesh used to endure a big flood every 20 years. Today there's a new pattern: a "1-in-20-years flood" hits every five or 10 years

user, it will also try to make itself resilient to future climate changes," says West. For example, it will build flood protection into its design. "This could be a new model for how communities can walk on both legs into the climate future."

Bangladesh

AS A LOW-LYING COUNTRY THAT FACES THE SEA AND DRAINS 92% of the snowmelt from the vast Himalayan mountain range, Bangladesh is one of the most vulnerable places on the earth to global warming. Already, sea levels are rising in the Bay of Bengal and pushing salty water inland, reducing the amount of land on which rice can be cultivated in the south of the country. Farmers are adapting by switching to farming prawn, which can tolerate saltier water.

"Bangladeshis have lived with flooding forever. It's part of our culture and essential to our agricultural system," says Saleemul Huq, who directs the climate-change program for the International Institute for Environment and Development. "In the past, we experienced a very big flood about once every 20 years," Huq says. "But in the last 20 years, we've had four very big floods—in 1987, 1988, 1995 and 2005. So it appears that the new pattern is to get a 1-in-20-years flood every five or 10 years." That increase has gotten policymakers' attention. After years of lobbying by Huq and his colleagues, the Ministry of Water Resources recently agreed to incorporate climate-change models into all future planning and decisions.

But because of its poverty—78% of its population lives on less than $2 a day—Bangladesh cannot afford the kind of de-

Fighting the Meltdown

As the globe heats up, the ramifications are felt from the Alps to the deserts of China. Scientists and engineers are developing creative new ways to "geoengineer" the landscape and compensate for dated technology that harms the planet's atmosphere. China, one of the world's most polluted nations, is just beginning to promote green science

Scouring powder *In 2006 Chinese weather specialists seeded clouds with chemicals to engineer artificial rainfall, hoping to relieve drought and cleanse smog and dust from Beijing*

Green belt *To reduce windstorms in Beijing, Chinese scientists are planting a "green belt" around the capital. Reforestation will be aided by conserving rainwater to increase soil fertility*

fenses planned in Europe, or even New Orleans. As a matter of fairness, Huq says, adaptation measures in poor countries should be subsidized by rich countries. "It is poor countries that are suffering the brunt of climate change," he says, "but it is the rich countries' greenhouse-gas emissions that caused this problem in the first place."

Britain is already subsidizing a substantial program in Bangladesh that will raise roads, wells and houses above the level of the last major flood. "Bangladesh is a showcase of what will happen under climate change," says Penny Davies, a diplomat

Snow job *As the world's roofline heats up, ski slopes formerly covered with snow for most of the year are melting down. Above, workers spread an insulating cover over the Tortin Glacier in Switzerland to preserve snow*

Cold storage *Inuit houses in Kotzebue, Alaska, use thermopiles —devices designed to help keep permafrost frozen—in order to prevent meltdown and soil erosion around foundations*

Obsolete *Old refrigerators taken from New Orleans after Hurricane Katrina are set aside so their cooling freon gas can be recaptured, rather than leak into the atmosphere. Chlorofluorocarbons in freon damage the planet's ozone layer; their use was banned internationally in 1989*

at the British High Commission in Dhaka. "It amounts to a testing ground for what island states, including Britain, will need to do to protect ourselves in the years ahead."

The U.S. has a long way to go before it is climateproof, but so does most of rest of the world. Japan has an impressive, long-standing system of flood control, including the so-called G-Cans project, a massive underground system in Tokyo that can pump 200 tons of water per second out of rivers and into the harbor before the city's streets flood. But former city officials acknowledge that Tokyo's system has reached its capaci-

ty. Since global warming is expected to bring Japan more frequent torrential rains, Tokyo will have to upgrade its drainage and sewage systems.

The latest science makes clear that we will be living with global warming for the rest of our lives. That's not a happy thought, but it's not necessarily a dire one either. The key is to follow the new rules of life under global warming. Think ahead, adapt as necessary and strongly support cutting greenhouse emissions in time. Adaptation won't be cheap. It won't be optional either.   ∎

Overhead charges *A 60,000-sq.-ft. array of solar panels on the roof, installed in 2004, helps light the interior of the Moscone Center in San Francisco, a large exhibition facility*

Power Plays: Searching for Clean Energy

Can a world hooked on hydrocarbons create alternate sources of power to replace our planet-damaging 19th century standbys, Old King Coal and polluting petroleum?

HERE'S A CHALLENGE: IMAGINE FOR A MOMENT THAT you live in a world that derives all its energy from a source of power that is about to be depleted. Suppose that the energy that has warmed our homes, brightened our nights and moved our goods for more than a century will shortly be unavailable. For added complexity, suppose that not everybody was convinced of this fact, although you know it to be true. The good news is that scientists are pretty sure that an entirely new source of energy can be tapped to meet these needs. Your job is to make this enormous, paradigm-shifting conversion a reality.

What would you have to do? The task would challenge you to engineer new technology to harvest the fuel, create new engines to turn it into energy and build new infrastructures to distribute it. And that's the easy part. Along the way, you would need to convince bankers, politicians and the public to abandon comfortable orthodoxies, embrace radical change and invest in unproven ideas. All because you say the world as we've always known it is about to disappear. Would you be greeted as a savior and visionary? Or would people call you a lunatic and laugh while they ignored you?

Ponder this question, and you will have some idea of the struggle being waged by people who believe that hydrocarbon power, today's chief energy source, is going to have to be phased out or radically cleaned up if the planet we live on is to continue to sustain life. Early on, these people were brushed off as extremists. Then they were derided as well-intentioned folk who simply weren't practical. Only in the past decade have decision makers and the general public begun to realize that new forms of energy are essential to our survival—and many of us have done so reluctantly, ambivalently and partially.

But the truth is, we've done this before. For centuries in Europe and early America, food was cooked, homes heated and evenings illuminated by burning wood, and later whale oil. It

port found that government incentives designed to encourage more widespread adoption of technology that already exists—combined with legal mandates to make cars, homes and factories more energy efficient, as well as aggressive investment in renewable energy—could hold global temperature increases to around 3.6° Fahrenheit above pre-industrial-era levels, which is low enough to avoid potentially disastrous droughts, severe storms and sea-level rise.

One of the frustrating myths about alternate energy is that we're still waiting to find a single energy source to replace hydrocarbons. The truth is, we may never do so, because no one source is likely to fit the bill. While oil and its cousins, coal and natural gas, are now king, they will most likely be succeeded not by another absolute monarch but instead by a parliament of many energy sources.

If the task of replacement seems daunting, it is helpful to recall that the people who built our hydrocarbon economy faced similar obstacles and in conquering them not only made the world a better place for everyone, but also enriched themselves in doing so.

In the years to come, fortunes will have to be invested on drilling wells into the ground (looking for heat instead of oil), laying vast networks of pipelines (to transport hydrogen rather than fossil fuels), building new power plants (that collect energy from the sun and wind) and manufacturing cars that run on electricity or on fuel derived from corn, sugarcane or hydrogen instead of hydrocarbons.

Benchwarmer *A windmill atop this British bus station heats its seats. In the future, more power will be generated onsite rather than created centrally and transmitted via power lines*

was only in the mid-1800s, after the forests had largely been felled and whales hunted to the brink of extinction, that people began listening to energy prophets like Edwin Drake, who believed that the black slime bubbling out of the ground in places like Oil Creek, Pa., could be used for fuel.

The revolution unleashed by Drake and other petroleum pioneers, like John D. Rockefeller, remade the world. We now find ourselves at a similar threshold, not because the world is quickly running out of hydrocarbons (although some experts argue persuasively that it is) but because we can't live much longer with the consequences of their continued use.

The earth is, quite simply, choking on greenhouse gases. Global carbon dioxide output in 2006 approached a staggering 32 billion tons, with about 25% of that amount coming from the U.S. Turning off the carbon spigot is essential, and many of the proposed alternatives are familiar: windmills, solar panels and nuclear plants. All these technologies are already part of today's energy mix, though each has serious drawbacks.

But there is reason to be encouraged. A May 2007 report produced by the United Nations Intergovernmental Panel on Climate Change concluded that many of the worst consequences predicted for global warming can still be averted by making the switch to renewable, nonpolluting sources of energy. The re-

Many of the worst effects of global warming can still be averted by making the switch to renewable, non-polluting sources of energy

The pioneers who succeed in doing so will be the Rockefellers of a new age, remaking and improving the world.

In the meantime, the phases by which we have come to acknowledge global warming and begun to relinquish our embrace of hydrocarbons closely track the five stages of grief experienced by terminally ill patients, as formulated by Elisabeth Kübler-Ross in the 1960s. First came Denial ("This isn't happening"), then Anger ("How can this be happening?"), followed by Bargaining ("Just let us use dirty energy for one more generation, then our children will switch to alternate sources of power"). The fourth step is Depression ("There's nothing we can do about this"). Finally, there is Acceptance ("The world that we're moving toward is a better one"). In the first decade of the 21st century, we appear to have just about finished Bargaining, and we're now facing the question of whether we can skip Depression and go straight to Acceptance. ∎

Solar Power: The Outlook Brightens

The good news is that solar power has never been hotter. More than 1 million rural homes in developing countries get electricity from the sun, and fully 80% of China's hot water comes from solar cells. The not so good news is that while solar power has enormous potential, it is decades away from supplying a significant slice of the world's overall energy consumption. As of 2006, according to the International Energy Agency, that share was 0.4%.

Why? To begin with, the technology is still in its infancy. The first practical photovoltaic cell, which uses silicon to convert sunlight into electricity, was developed in 1954, making solar power younger than the transistor. Yet solar has lagged, while the latter technology has already transformed the world.

Perhaps more important, solar power has an image problem. Since their invention, solar cells have proved far more useful in generating publicity than electricity. Decades of inflated hopes have been followed by deflating results. The last time many Americans paid attention to solar power was in the 1970s, when a series of oil shocks prompted President Jimmy Carter to install solar panels on the roof of the White House. (They were removed by the Reagan Administration a few years later.) In Carter's day, solar equipment was bulky, ugly, anemic in output and very expensive, a recipe for public resistance.

Most people remain in the dark about solar power's capabilities. One persistent myth: solar cells don't work on cloudy days. In fact, photovoltaics will generate electricity in any kind of day-light, although their output is greatest when the sun is brightest. Few people realize that solar-power technology has evolved several generations since the 1970s. Solar cells now convert as much as 17% of the energy they gather from the sun into electricity, up from 6% when they were developed. At the same time, the price of cells has plummeted: solar cells cost more than $200 per watt of generating capacity in the 1950s; the figure was $2.70 per watt in 2004. Since then, a silicon shortage has pushed the price back up, to approximately $4 per watt.

Despite the silicon shortfall, the technology is steadily improving. One vision of the future of solar power can be found in the desert outside Las Vegas, where the $250 million Nevada Solar One power plant is scheduled to come online in the summer of 2007. This state-of-the-art facility will use acres of mirrors to focus solar radiation onto a turbine, producing 64 megawatts of current—enough to power more than 40,000 homes—more cheaply than almost any other solar plant in the world.

The U.S. Department of Energy has set a goal of making solar power cost-competitive with more traditional forms of power, like coal and oil, by 2015. Department scientists calculate that if solar panels covered just 0.5% of America's landmass, they could generate all the electricity the nation currently requires. At least for now, however, the maddening conundrum remains: a clean source of power that falls from the sky, cuts out the middleman and for which the fuel is free still costs more than the dirty old combustibles we have to pull out of the ground.

Sign of the times *Workers install a solar panel on a home in rural India, outside the town of Hanuman Nagar*

■ Wind Power: Another Revolution?

Supporters of harvesting energy from moving air can no longer be accused of tilting at windmills. Over the past three decades, the cost of a watt of electricity generated by the breeze has dropped more than 80%, meaning that powering your air conditioner from a wind farm now costs roughly the same as hooking it up to a coal plant—and with no carbon emissions. The biggest innovations have been economies of scale: in wind, it turns out, bigger usually is better. During the 1970s, the last time the search for alternative energy made headlines, turbine blades 32 ft. long were considered state-of-the-art, and wind power cost $2 per killowatt hour (kW-hr). Today, wind turbines with blades 130 ft. long have cut the cost to around 5¢ per kW-hr. after a federal subsidy kicks in. That's just a penny more than power from coal-fired plants.

What have we learned? Bigger turbines are inherently more efficient because more wind is caught by a blade with a larger surface area. So a town of 8,000 needed 40 turbines to meet its power needs in the mid-1990s, while just one larger model can do the trick today. On the drawing board are turbines with blades almost 300 ft. across—at which point, many experts say, wind power may start to cost less than hydrocarbon power.

No wonder power companies and governments around the world are anticipating, well, a windfall. Denmark, the world's No. 1 builder of wind turbines, already pulls 20% of its electricity from the air, while Germany, the global front runner in producing electricity from wind, generates about 5% of its power from the wind. In the U.S., less than one half of 1% of all electric power comes from the breeze, but that federal tax subsidy of 1.8¢ per kW-hr has made wind power (along with solar) one of the nation's two fastest-growing sources of electricity. As a result, U.S. wind capacity more than doubled between 2000 and 2005. Look for more growth in the years ahead.

Turn, turn, turn *A battery of wind turbines forms a curving snake across the landscape of Alameda County, Calif.*

Mercury rising *A geothermal plant in Iceland is located near hot springs warmed by heat from the core of the planet. Rubber duckies are optional*

Geothermal Power: The Inside Story

It's one of geology's immutable laws: the deeper you dig, the hotter it gets. Because the center of the earth is hotter than the surface of the sun, nearly endless supplies of energy in the form of heat lie just a few miles beneath our feet. But how do we tap it? Geothermal energy today is roughly where petroleum energy was in the 1870s: it's an infant industry, scooping up the cheap-and-easy heat lying close to the surface rather than drilling down to reach the prime supplies. Even so, geothermal is bigger than most people realize: it now supplies electricity (mostly in the West, where heat can be found not far below ground) to 2.8 million U.S. homes, more than solar and wind sources combined.

But all this may be just a hint of what's to come. A 2007 report from the Massachusetts Institute of Technology calculated that a government investment of $500 million to $800 million over the next 10 years to develop a new technique called "heat mining" could yield a capacity of 100,000 MW (roughly the output of 200 large coal-burning power plants) by 2050. Instead of being confined to regions where warmth reaches the surface, heat mining involves looking for hot rocks deep underground and can be done in many places.

Heat mining could replace a generation of aging coal and nuclear plants that are scheduled to go offline during the next 25 years, and with zero greenhouse-gas emissions. But before that happens, geothermal will have to survive 2008, a year in which the 1.9¢-per-kW-hr tax credit that has supported geothermal development since 2005 is set to expire, and in which the U.S. Department of Energy has proposed killing funding for geothermal research.

Gasification: Coal Comes Clean

Not all forms of alternative energy involve new sources of power; some of them focus on cleaning up dirty old ones. Transforming solid coal into a vaporous gas is nothing new: streetlights in many 19th century U.S. towns were powered by "town gas," which was extracted from coal locally. Today this old trick has taken on a new, clean twist: coal gasification involves baking coal at very high temperatures while depriving it of oxygen so that it can't burn. The coal releases a synthetic form of natural gas (syngas) and leaves most of the CO_2 behind.

Gasifying coal is more expensive than burning it but sharply cuts greenhouse-gas production and still taps America's abundant coal resources. Florida's Tampa Electric Co. has churned out electricity reliably for 10 years with a coal-to-gas system that removes pollutants before combustion, reducing emissions.

Pipe cleaner *A Louisiana plant produces "clean coal" synthetic gas*

Ethanol Power: Putting Corn in Context

Turning plants into fuel is a growth industry. And no biofuel is coming on stronger than ethanol, which is another name for pure alcohol. Chemists and moonshiners have long known how to reduce crops like corn and sugarcane to alcohol. But it was President George W. Bush's 2006 State of the Union address that put the business into high gear. Bush, a former Texas oil-industry man, called for the U.S. to quintuple its production of biofuels, primarily ethanol, by 2017. America has some catching up to do: Brazil, which committed to ethanol decades ago, meets 40% of its transportation needs by filling its fuel tanks with ethanol distilled from sugarcane.

What's the attraction? As well as burning much more cleanly than ordinary gasoline, ethanol obviates the need for a widely used gas additive, a toxic substance called methyl tertiary-butyl ether (MTBE), that helps car engines run more smoothly and pollute less. But MTBE also pollutes groundwater, and since 2003, California has been replacing MTBE with ethanol by law.

In 2006, more than 2.1 billion bushels of U.S. corn were turned into ethanol. That's twice the level of 2002 and more than one-fifth the entire U.S. crop. As a result, despite a bumper 2006 harvest, corn sold for a 10-year high of well over $3 a bushel, raising the price of dairy, poultry and pork products, all of which rely on corn for food. More than 3 billion bushels of corn are expected to be used for ethanol by the end of 2007.

Corn-based ethanol is far from a perfect replacement for gasoline, however—though your chances of hearing that at an Iowa political caucus are probably slim, since the ethanol band-wagon runs on the contributions of farming interest groups. The current inefficient manufacturing process that turns corn into ethanol burns up almost as much energy as it produces.

Extracting fuel from sugar, as in Brazil's case, makes more sense; this process yields eight times the energy it consumes. But the U.S. grows relatively little sugarcane, and high import tariffs discourage importing more. Better still would be to process ethanol from agricultural waste like wood chips or the humble summer grass called switchgrass, a formerly obscure plant Bush put into the national vernacular in the same State of the Union address. This "cellulosic" ethanol packs more energy than corn ethanol, but it also takes more energy to produce.

Another strike against ethanol: all varieties of the substance cost more than ordinary gasoline, at least for now, and the use of ethanol as a fuel remains financially viable only because of a 51¢-per-gal. tax exemption granted by the Federal Government to refiners who produce a gasoline-ethanol blend. Critics often cite another problem with ethanol: it currently must be transported to refining plants by old-fashioned trucks and trains, burning emissions-producing hydrocarbons in transit.

Ethanol isn't a magic bullet for America's fuel problem; it's a temporary fix. Most U.S. cars now run on a gasoline-ethanol mix that in many cases includes no more than 10% ethanol. Large-scale consumption would require converting to a fleet of flex-fuel vehicles that can use gasoline containing 85% alcohol. The differences in engine design are relatively minor, but over-hauling America's entire fleet of vehicles is a daunting prospect.

By some estimates, even if such an overhaul were to occur, converting the entire U.S. corn crop to fuel would offset only about 12% of America's gasoline consumption. And, as Bush said not long after his 2006 State of the Union address, "you've got to recognize there are limits to how much corn can be used for ethanol. I mean, after all, we got to eat some."

The reign of corn *Not surprisingly, the federal subsidy on corn-based ethanol is strongly supported by Midwestern politicians*

Nuclear Power: The Fallout

For many people, nuclear power still evokes two images: Three Mile Island and Chernobyl. But proponents of green energy are taking a fresh new look at it. The main concerns are the disposal of nuclear waste, which can remain radioactive for hundreds of thousands of years, and the safety of nuclear reactors. Nuclear's advocates note that a typical reactor produces around 30 tons of dangerous waste a year, which can theoretically be safely stored underground, while the average coal plant produces 1,000 tons of harmful emissions each day. Nuclear's foes say, Note that *theoretically*.

The safety issue may be overblown. The world has 440 nuclear power stations, some of which have been in operation for 50 years. In that time there have been two dozen accidents, with only one major calamity, at Chernobyl. Nuclear plants are the Energizer Bunnies of U.S. power generation: they are not only reliable and long-lived, but they also create more juice year after year, thanks to technological advances. All told, nuclear plants now produce some 20% of America's electric power, up from just 4.5% in 1973.

A new wave of modular (purportedly even safer) reactors is coming online; their builders say they will never need to be shut down for refueling and will be practically meltdown-proof. The kicker: nuclear power is unlike most other alternate energy sources in one crucial respect—it works now.

Core corps
Workers at a nuclear plant in Germany

Hydrogen Power: Ready for Prime Time?

Americans concerned about both energy and the environment were inspired when the White House announced a new program to develop "hydrogen-fueled vehicles … to enable a shift away from oil," along with a firm target date to begin the transition. But the President was Richard Nixon, the speech was given in 1974, and the date for beginning the switch was 1990. (For the record, George W. Bush, in his 2003 State of the Union address, also promised "that America can lead the world in developing clean, hydrogen-powered automobiles.")

Hydrogen is the most abundant element in the universe, and it burns far more cleanly than fossil fuels. So why are we all still choking on what comes out of our tailpipes?

Hydrogen hides in plain sight. There are bountiful supplies of it everywhere we look, but it is almost always chemically locked in compounds like water, which weaves hydrogen together with oxygen and is tricky to undo. Ironically, our best current way to get power from hydrogen is by burning oil, coal and natural gas. Their concentrated hydrogen content is what gives them energy in the first place. It's the carbon part of "hydrocarbon" that causes all the problems.

Today, running a car on hydrogen without using carbon involves using either hydrogen fuel cells or ordinary engines modified to burn hydrogen. The first is perhaps the oldest cutting-edge technology we're still trying to perfect. Invented more than 150 years ago, the fuel cells combine hydrogen and oxygen, producing heat and water. The heat is used to create electricity; the water is a waste product. But fuel cells are still too pricey to be widely used.

What about Plan B? Retooling a car's engine to run on hydrogen is both simple and inexpensive. But obtaining pure hydrogen by extracting it from its molecular partners currently uses up more energy than the resulting hydrogen yields. So for now, hydrogen's future seems to belong to fuel cells, whose cost, if high, has dropped sharply in recent decades. New technologies and economies of scale should make the cells affordable in the not too distant future. But for now, 1990 is still a few years away.

compressed **700**

Running clean *Gassing up … or, uh, hydrogening up, at a station in Washington*

■ Pioneers of Alternate Energy

Angela Belcher
An M.I.T. polymath uses genetic engineering to make a better car battery out of viruses

The reason we aren't all driving electric cars has little to do with a Detroit conspiracy. It's that nobody has invented a lightweight, inexpensive battery that can store enough electricity to make such a vehicle practical.

If anyone can change that, it's Angela Belcher. A materials scientist and bioengineer at M.I.T., Belcher, 49, won a Mac-Arthur Foundation "genius" grant in 2004; in the fall of 2006, *Scientific American* named her research leader of the year for her current project: creating an entirely new kind of battery, not by building it but by growing it. Working with several M.I.T. colleagues, Belcher has engineered a virus, known as an M13 bacteriophage, that latches onto and coats itself with bits of inorganic materials, including gold and cobalt oxide. That turns each long, tubular virus into what amounts to a minuscule length of wire. Coax these nanowires to line up, and you have the components of a battery that is far more compact and powerful than anything available.

If her battery works as a commercially viable product, that alone could qualify Belcher as a climate-change hero, but her vision is green in other ways as well. Conventional batteries generate a lot of waste during manufacture, and they're a disposal nightmare. But a viral battery essentially grows itself, using water as a solvent, so there's practically no waste. And since much of its relatively small bulk is organic, the battery is partially biodegradable. Goodness in, goodness out.

Robert Socolow and Stephen Pacala
These Princeton profs propose we limit our carbon emissions with a seven-step program

While the solution to global warming seems dauntingly complex, physicist Robert Socolow, on left, and ecologist Stephen Pacala have come up with a remarkably straight-forward way of approaching it. To stabilize the world's carbon emissions, they propose not chasing a single magic bullet but harnessing seven different categories of reduction, using available technology. Their goal is to draw a road map for reducing CO_2 emissions that is both realistic and effective.

Each of the strategies they have identified could prevent a total of 25 billion tons of emissions by 2056. (We're now adding 7 billion tons annually, and that figure would double by 2056 without some action.) They are all so-called stabilization wedges, which lower the angle of the rising line representing carbon-emissions growth and together would reduce CO_2 emissions enough to stabilize the level of carbon concentration in the atmosphere. Efforts to reduce energy use form one kind of wedge. So does improving power plants. Another wedge addresses alternative energy, and so on.

As co-directors of the Carbon Mitigation Initiative at Princeton University, Socolow and Pacala oversee research exploring the potential of wind, solar, hydrogen, geothermal and, yes, even nuclear power to contribute to several of the wedges. The stabilization wedge may prove to be a useful way of thinking about other vectors in which science and policy intersect. The concept has earned Socolow a seat on a National Academy of Engineering panel that will try to figure out the greatest engineering challenges of the next 100 years.

Shuji Nakamura

Working in obscurity, a researcher creates a superefficient lightbulb for the 21st century

The world is a brighter place because Shuji Nakamura is not easily discouraged. In 1993 he astonished the scientific community by creating the first successful blue light-emitting diode, or LED. The blue LED was the last step in achieving lighting's holy grail, the brilliant white LED—an ultra-efficient successor to Thomas Edison's incandescent bulb.

At the time of his breakthrough, Nakamura was an unknown engineer without a Ph.D. working for a tiny Japanese electronics company. After colleagues told him he had wasted research money for 10 years, he decided to follow his instincts. He did, and the blue LED was soon followed with the first bright green and white LEDs. An LED is a semiconductor that generates light, but very little heat, when an electric current is passed through it. Different semiconductors produce different colors; Nakamura used gallium nitride, which generates blue and white light.

The resulting LEDs use as little as one-seventh the energy as an incandescent bulb and can last about 100 times as long, up to 100,000 hours. If they were widely used, LEDs could lead to enormous energy savings and carbon-emissions reductions. In the developing world, LEDs paired with solar panels could provide a cheap, sustainable light source that doesn't need a traditional power grid.

Nakamura, now at the University of California at Santa Barbara, won the $1.3 million Millennium Technology Prize in 2006 for his work on LEDs. He is now researching zero-energy-loss LEDs, which would be close to 100% efficient.

Fred Krupp

This power broker lobbies companies to use cleaner energy by rewarding good behavior

Fred Krupp, the president of lobbying group Environmental Defense, is calling on Congress to approve a system that would mandate reductions in CO_2 emissions and allow the sale of permits to release specified amounts of carbon.

Under his system, companies having trouble cutting their emissions could buy allowances from firms that have unused permits. Or they could pay farmers to store carbon and reward developing nations for preserving forests.

The idea comes from a concept developed by Environmental Defense when Krupp helped draft the 1990 Clean Air Act. It set up a trading system to control sulfur dioxide. Krupp believes similar financial incentives could slow global warming. "Once you put a value on carbon reductions," he says, "you make winners out of innovators. You offer a pot of gold."

Building Green

A new design aesthetic is sweeping the world, as builders, engineers and architects rethink structures for a warming globe. The message: Don't fight nature—make it a partner

THE ADAM J. LEWIS CENTER FOR ENVIRONMENTAL Studies at Oberlin College, above, is a beauty of a building. But if the structure's façade resembles a jewelbox reflected in a watery mirror, it's what's inside the building that makes it so compelling to envious architects: the state-of-the-art disinfectant system that

cleans toilet water for reuse. (No, not in the drinking fountains.)

Now consider the Philip J. Merrill Environmental Center in Annapolis, Md. It's as earth-friendly as an old windmill. The headquarters of the Chesapeake Bay Foundation, it displays more wood construction than the typical large building these days. But to understand what its designers did to make it truly different, you would have to know that one-third of the energy it uses comes from geothermal heat pumps that utilize the earth's warmth and photovoltaic building panels that convert sunlight into electricity. Or that rainfall collected on the roof can be channeled into huge holding tanks for reuse in irrigation. Or that its sunscreen overhangs are made from recycled pickle barrels. Whole platoons of enforcement lawyers for the Environmental Protection Agency could not be more ecologically effective than its waterless composting toilets, bamboo floors and timber cut from sustainably harvested wood.

These buildings epitomize today's green architecture, a catch-all term for design and construction practices that take into ac-

count a whole checklist of environmental goals. How a building is sited, how well it reuses its wastewater, how efficiently it is heated and cooled—those are all questions green architects examine closely. To answer them, they employ a new generation of supplies that include nonpolluting paints, low-flow toilets and windows glazed to admit sunlight but reduce heat radiation. But green design is not all about high tech. One simple idea: windows on high-rises that actually open. That facilitates natural air-ventilation systems, also known as breezes. Eureka!

The thing about buildings is that they are, par excellence, the very thing nature is not. Ever since people moved out of caves, which were pretty much all natural if you don't count the

Shipshape *Old metal shipping containers are piled together to form a colorful recycled house in London, complete with nautical portholes*

paintings on the walls, structures have been the prime markers of human settlement, a process that often comes with unhappy consequences for the environment. John Denver's *Rocky Mountain High*—"More people, more scars upon the land"—is not a song you hear much at architecture conventions.

No one can deny that when it comes to the environment, buildings are right up there with automobiles as polluters. Homes, schools, office towers and shopping centers dirty their own little rivers of water every day. Their air-conditioning and heating systems waste large amounts of electrical and fossil-fuel power. Toxic ingredients leach from building materials and foul the air. Thirty years ago, only a few environmentally minded architects cared about such things.

That began to change in the 1970s with that decade's oil shocks, which produced a short-lived vogue for alternative heating technologies. The simultaneous rise of environmentalism also inspired what you might call hobbit architecture, cottages crowned with listless greenery and the odd solar panel. But it wasn't until the 1990s that green architecture gained a foothold in mainstream building. That was in part the result of a growing realization that "sustainable" buildings have lower long-term heating and cooling costs. States began offering tax incentives for construction that put less pressure on power grids or water supplies. Coming of age at the same time was a generation of architects who were knowledgeable about environmentally conscious construction materials and techniques.

In 1998 the U.S. Green Building Council, an association of architects, builders and other green specialists, adopted the Leadership in Energy and Environmental Design (LEED) certification system, which sets out standards that a building must meet to qualify as environmentally friendly. The council estimates that some 3% of new building starts each year have a few earth-friendly features, and the number is increasing. "The growth of green building is driven partly by energy efficiency and other cost savings," council president and CEO Christine Ervin told TIME in 2002, "but also by the need of businesses to attract the best employees. These buildings can make very attractive workplaces."

Some of the most prominent names in architecture have turned green. The three-sided Commerzbank Tower in Frankfurt, Germany, is a major work by a renowned British architect, Sir Norman Foster. At 53 stories, it is not only one of the tallest buildings in Europe but also one of the leafiest. All around its triangular interior atrium are gardens in the sky, set at different elevations, so that no worker is more than a few floors away from a sizable patch of greenery, visible through windows that actually open.

Natural air circulation is a preoccupation of green architects. With the widespread adoption of air conditioning after World War II, office buildings were built to be more airtight than a mummy's tomb. Now designers are rediscovering principles of ventilation and air circulation familiar to 19th century builders. The Rocky Mountain Institute took part in an environmental upgrade of the the vintage Executive Office Building and the White House in 1993. "We discovered that the old office building was already designed with a natural ventilation system—a fairly brilliant one," says William Browning, the institute's senior consultant for green development. Parts of that system, which once linked chimneys and other air passages, are back in operation.

Not everything green is rosy. To provide sunlight that reduces reliance on electrical lighting, environmentally conscious designers tend to favor open-plan workplaces over offices with doors that close. That can be good for nature, less good for quiet and privacy. And big suburban residential developers are not piling in yet. Reduced long-term energy costs, for instance, are not an important incentive to builders who plan to sell right away the homes they build.

Some green architecture is literally green. Dwellings that nestle directly into the landscape like caves, with carpets of

When Blueprints Go Green

The U.S. Green Building Council's Leadership in Energy and Environmental Design (LEED) program measures a building's planet-friendly factor based on five criteria:

- Sustainable site development
- Water savings
- Energy efficiency
- Materials selection
- Indoor environmental quality

earth and grass rolling over them as roofing, were among the first examples of green architecture in the 1970s. Buildings like those may be related to Bronze Age settlements dug into the earth. But they operate on principles that can be adapted to modern midtown high-rises. Since 2001, Chicago's City Hall, a 1911 Classical Revival building, has been topped by a "green roof": a 20,000-sq.-ft. garden that was planted as a climate-control mechanism. In summer the garden helps keep the building cool by shielding it under a layer of moist material. In winter it insulates against cold. Not incidentally, it also provides a habitat for birds, butterflies and grasshoppers. But not yet for people—the garden is closed to the public. Sometimes nature needs to work in peace. ∎

Philip J. Merrill Environmental Center

Architect Tom Eichbaum, SmithGroup • **Location** Annapolis, Md. • **Year Completed** 2001

The headquarters of the Chesapeake Bay Foundation was created to be a showplace for green design, and it has the certificate to prove it: it was the first structure to receive a platinum rating under the U.S. Green Building Council's LEED program. The foundation, committed to preserving the great Atlantic bay, boasts that the center was built on a "cradle to cradle" philosophy: all construction materials were recycled or created through processes that do not harm the environment, and they are intended to be recycled again when they wear out.

The structure was completed in 2001, and visitors will find a number of features that are still percolating through the world of architecture. The windows open. Plants are native. Bike racks and plug-ins for hybrid vehicles are plentiful. "Heat islands"—spaces such as asphalt parking lots and black rooftops that trap high temperatures—are minimized. (The center's parking lots are gravel.) The building uses both passive and active solar energy; photovoltaic panels are located on the roof. It also draws geothermal energy from wells drilled below the frost line, where a constant temperature of around 50°F helps cool the building in summer and heat it in winter. All told, the Merrill Center meets one-third of its energy needs from nontraditional sources.

The building's toilets are waterless composting units that convert human waste into garden soil over a period of three years. All other water needs are supplied by retained rainwater, captured in recycled barrels from a local pickle factory, which also serve as window canopies, screening out the sun.

Magney House

Architect Glenn Murcutt • **Location** Bingie Bingie, New South Wales, Australia • **Year Completed** 1984

Aesthetics meet ecology in the buildings of Pritzker Prize–winning Australian architect Glenn Murcutt. He designed Magney House, which overlooks the ocean, for a couple who had camped on the rugged site for years and wanted a home that would remind them of a tent. The sloping roof echoes the contours of nearby hills and catches rainwater, used both for drinking and to cool the house. Louvers, eaves and sliding doors protect the home from the scorching sun and bone-chilling cold. An undulating ceiling redirects natural light to every area of the house. Metal sheathing, brick walls and an insulated foundation slab lock heat inside during winter and keep it out in summer, conserving energy year-round.

Commerzbank Tower

Architect Norman Foster • **Location** Frankfurt, Germany • **Year Completed** 1997

Green architecture went mainstream in 1997, with British architect Sir Norman Foster's design for the Commerzbank Tower in Frankfurt, Germany. At the urging of Germany's Green Party, which governed Frankfurt in the early 1990s, when the building was being planned, Foster included many environmental innovations, such as high-performance glass that deflects heat and reduces the need for air conditioning.

While generations of Modernist architects used technical tricks to create the illusions of "air" and "light" within sealed environments, Foster went low-tech to let in the real things. The windows on Commerzbank Tower actually open, a rarity in high-rise office buildings. And the three-sided building is hollow—a triangular atrium soars more than 700 ft. from the lobby to the building's roof, further helping air circulate throughout the structure.

The atrium is connected to the outside world by nine three-story "sky gardens," which spiral around the building's exterior at eight-story intervals. The spacing means that no point within the building is more than a few floors away from a green space, and almost every room in the structure has a view of at least one garden. The large openings in the building's skin created by these gardens allow fresh air to ventilate through the entire tower, while also permitting natural light to flood the interior. Throughout the day, most offices receive at least some direct sunlight, either from the outside or from daylight that shines through the gardens and the atrium. Not only does this make for a pleasant working environment, but the design requires less electricity for artificial light. Electric consumption is cut further by automatic dimmers, which sense natural light and turn down lamps.

Some of the building's most important design improvements are hidden behind walls and ceilings. Much of the open space is made possible by the use of steel as the building's main construction material, rather than cheaper (and bulkier) concrete. Internal louvers boost air circulation. Water consumption is reduced by the use of "gray water" (which has already been cycled through the climate-control system) to flush toilets. The sum of all these innovations is what Foster calls "the world's first ecological office tower." The influential architect, knighted in 1990 and made a life peer in 1999, has gone on to create other planet-friendly office buildings around the world, including the Hearst Tower in Manhattan.

SIEEB Building, Tsinghua University

Architect **Mario Cucinella** • Location **Beijing** •
Year Completed **2006**

Designed by Italian architect Mario Cucinella and built by a partnership of Italian and Chinese construction firms, the Sino-Italian Ecological and Energy-Efficient Building (SIEEB) at Tsinghua University in Beijing was commissioned to be a showcase for green building technologies and a model for future designs that aim to reduce CO_2 emissions in China, one of the world's leading producers of greenhouse gases.

The 10-story, U-shaped structure is designed to be a magnet for sunlight: its two wings feature setbacks that allow the sun to filter through ceilings and to be gathered by the solar-power panels visible at the end of each level. The building also features adjustable canopies that extend or retract automatically, based on the season, temperature and time of day. As a result, it absorbs maximum sunlight in winter and deflects most solar heat in summer.

The building is powered entirely by natural gas, which supplies both heat and light. It is designed to consume 70% less energy than other Chinese buildings of similar size. Radiant heat sources are located in ceilings; the heat is reduced when sensors detect a room is empty. Water is recycled within the building—a hallmark of green design— and a computer-operated intelligent building management system automatically switches off lights and climate-control systems when they are not needed. Architect Cucinella was honored in 2004 as the year's Outstanding Architect by the U.S.-based World Renewable Energy Congress.

Chicago City Hall Roof Garden

Landscape Architect **Weston Solutions, Inc.** • Location **Chicago** •
Year Completed **2001**

Chicago's Mayor Richard M. Daley put a green roof on the city's 1911 City Hall building in 2001. The 20,000-sq.-ft. garden features bottlebrush grasses, wild rye and thousands of other plantings. Created as a pilot program to reduce the building's cooling and heating costs and serve as a model for other Chicago buildings, it has succeeded in both aims. The insulating layer of soil and plants cools the building in summer and helps hold heat in winter, reducing climate-control costs by some $6,000 a year. And the idea has taken root on more than 300 other Chicago rooftops, making Carl Sandburg's City of Broad Shoulders the City of Aerial Gardens.

Caltrans Headquarters

- Architect **Thom Mayne, Morphosis**
- Location **Los Angeles**
- Year Completed **2004**

The Caltrans building in downtown Los Angeles is the District 7 headquarters of the California Department of Transportation, and fittingly enough, it's always on the move. One façade of the structure's 13-story main tower is covered in a mechanical skin that is often in motion, opening or closing depending on the temperature and position of the sun. The skin consists of perforated, moving aluminum panels over a double layer of glass. At noon the structure is closed up tight, shielding workers from the sun; at twilight it is transparent, welcoming the sunset.

The building's south façade is covered with photovoltaic cells that generate some 5% of its power needs. Where the solar cells meet the courtyard at the building's entrance, they form a canopy, a mini solar shield. Inside the building, elevators running on an energy-saving "skip-stop" basis run to mini-lobbies located every three floors, which offer welcoming gathering spaces. Interior stairways leading from the lobbies encourage workers to walk between floors. Yet however green the building, some Angelenos call its overall look hulking and forbidding.

Greenwich Millennium Village

- Architect **Ralph Erskine**
- Location **London, England**
- Year Completed **2003**

Architect Ralph Erskine is often described as Anglo-Swedish, for most of his career was spent in Sweden, and his designs reflect the angular modernism of Scandinavia, along with its emphasis on the use of natural materials. Erskine set a goal for the project, which is in London's Canary Wharf area, of reducing energy use 80% and water use 30%. A gray water system uses rainwater to flush toilets. As much as 40% of the wood and aluminum waste created during construction was recycled. Windows and roof canopies feature cedar louvers that act as sunshades and windbreaks.

The village, whose first section was completed in 2000, is on its own grid: it uses a combined heat-and-power system to provide central heating, hot water and electricity. Covering 72 acres, it will eventually consist of 1,079 apartments and 298 houses centered on an artificial lake and a village green (this is England, after all). The village will include shops and restaurants, a community center, primary school and health center. Not all the reviews are laudatory: some Britons, sensing a case of much ado about nothing, have charged that the village's green factor is more hot air than high tech.

▪ Hearst Tower

- Architect **Norman Foster**
- Location **New York City**
- Year Completed **2006**

British architect Sir Norman Foster, whose Commerzbank building is also shown in this story, is a pioneer of large-scale green building. His office towers configure space in new ways that give workers more access to light, air and one another. The Hearst Tower is a 46-story notched glass skyscraper covered with a webwork of triangles, called a diagrid, in off-white stainless steel. That serpentine frame uses 20% less steel than a conventional frame does. It's also delightful, converting the tower's exterior into a cage where sunlight plays all day. And because the diagrid divides the building into four-story segments, it provides a human scale that an unbroken glass-curtain wall would not.

The highlight of the tower is a massive indoor piazza, a 10-story atrium bathed in sunlight from overhead skylights that opens up the entire interior. The tower is green in one more way: it's an example of adaptive reuse, rising from the masonry walls of a six-story structure built as a headquarters for press tycoon William Randolph Hearst in the 1920s.

■ Pioneers of Green Building

Gail Vitorri

She's helping create standards that will make hospitals healthier for patients and the planet

With an estimated $200 billion in U.S. health-care construction planned for the next decade, hospital design and building will have a huge impact on the environment. And Gail Vittori means to have an impact on those hospitals. With her husband Pliny Fisk III, Vittori is co-director of the Center for Maximum Potential Building Systems, a nonprofit design center in Austin, Texas. MaxPot, as it's known, advises institutions of all kinds—from a homeless shelter in Austin to the Pentagon as it rebuilt after Sept. 11—on how to adopt environmentally sound materials and practices. But Vittori and Fisk have a special focus on health care. Two years ago, Vittori led a committee that devised *The Green Guide for Health Care,* a 360-page "design tool kit" that suggests steps that hospitals and other facilities can take to adopt green practices.

The guide, which can be downloaded at *www.gghc.org,* is currently the basis for more than 100 pilot projects across the U.S., including MaxPot's latest showcase: the new 500,000-sq.-ft. Dell Children's Medical Center of Central Texas, which opened in June 2007 and which aims to be the first hospital in the world to attain platinum LEED status. Its green features include heavy use of local and renewable materials, on-site wastewater facilities and windows that open. Its low-rise buildings are linked by courtyards, quiet spaces and gardens that reflect the plant life of the surrounding area. And as everybody knows, no color is more comforting than green.

Steven Strong

A veteran of America's first energy craze is working to bring down the cost of solar power for homes so that more of us can afford it

He used to be a "solar zealot." Those are Steven Strong's woods. "Even back in the solar-crazy '70s," he says, "it was an open question if anyone could survive trying to sell houses that produce all the power they need through renewable energy sources."

It wasn't exactly a smooth ride, but survive he did. Along the way, Strong, 56, whose firm, Solar Design Associates, is based in Harvard, Mass., turned himself into one of the nation's foremost experts on solar buildings. His initial breakthrough came in 1980, when he found a manufacturer to build his "integrated" solar roof, like the one that tops off the more recent barn at right. The first of its kind, Strong's integrated roof provided an alternative to the costlier—and clunkier—solar panels that are just slapped onto rooftops.

Strong is very aware that solar can increase the price of building a house about 15%. One way to push down cost is through economies of scale, which is why he's serving as consultant for the Sonoma Mountain Village Project, planned by California developer Codding Enterprises and scheduled for groundbreaking in 2008. "For the first time," explains Strong, "a developer has set course to create an entire town built according to principles of sustainability while keeping it competitively priced." Located 45 minutes north of San Francisco, the project is an entire community of environmentally conscious—and solar powered—apartments, houses and stores. "This," says Strong, "is solar for the people."

Natasha and Jeeth Iype
In India, new ideas on building green settle in with ancient social practices

Bangalore may be India's high-tech heart, but in one part of its leafy suburbs, there's a group of environmentalists trying to get back to the garden. In 2003, husband-and-wife architects Jeeth and Natasha Iype, working with Stanley George, a civil engineer, designed the Good Earth Orchard homes. Each of the 60 projected houses, now in various stages of construction, will feature slate and wood left in a natural state, without toxic waxes and finishes. Sewage will be treated in tanks that process waste without harmful chemicals. Household water will be heated by solar panels, which is expected to reduce electricity use—and electricity bills—30%. And whenever possible, local building materials are used, which reduces the need for gas-guzzling trucks to transport things from far away.

But the subtlest eco-friendly feature may be the verandas that open from each house onto a large, grassy courtyard shared by the entire community. The hope is that the shared space will encourage shared environmental awareness. "Building green homes is easy," says Jeeth. "Building green communities is incredibly difficult. You have to persuade a group of individuals to buy into the same ideologies."

Most Indian homes still house several generations. But as the country grows richer, more people are moving into Western-style single-family homes, using more energy and resources per family. The Good Earth Orchard team hopes to provide India's élite with green homes that meet their rising standards but provide space that is still in some ways shared. "Communities make sense in India, given our history of joint-family living," says Natasha. "We are simply trying to re-create what we had."

Enrico Borgarello
An Italian visionary turns ordinary cement into an emission-gobbling superadhesive

As head of research and development for Italcementi, Enrico Borgarello knows cement isn't the greenest of products. But under his direction, the Bergamo-based Italian company has developed a substance that could turn an ordinary building into a weapon against air pollution. It's called TX Active, and it's an additive for cement that literally eats surrounding smog.

"When light shines on TX, the material becomes active and neutralizes surrounding pollutants like nitrous oxide and sulfur dioxide," says Borgarello, who claims TX can reduce local air pollutants from 20% to 70%, depending on sunlight levels and wind. (It also adds as much as 20% to the cost of the cement.) Cover 15% of the exposed surfaces of a city like Milan, he estimates, and you could cut pollution in half. And TX helps buildings stay whiter than white by resisting pollutants that scar and stain concrete over time.

Can We Clean Up Our Act?

Seizing the initiative, leaders and politicians, companies and utilities, cities and towns—and even entire nations—are taking positive action to respond to global warming

SEATTLE MAYOR GREG NICKELS HAS NEWS FOR PRESident George W. Bush: global warming is also "local" warming. So for Nickels and his constituents, climate change is about the Cascade Mountains, where the city gets its drinking water and hydropower and where the snowpack has shrunk by half over the past 50 years. It's about the effect of Puget Sound's warmer waters on salmon runs. It's about hotter summers cooking up more smog. It's about a rise in sea level that could flood Seattle's port. "The stakes are high—globally and locally," he says. "We need to act."

Like Nickels, leaders around the world—in giant corporations and small mom-and-pop shops, in nations like Iceland and Sweden, in state and provincial governments—are coming together to address the challenges of climate change, forming a network of green action that girdles the globe.

In February 2005, when the Kyoto Protocol took effect in 141 countries but not the U.S., Nickels launched the U.S. Mayors' Climate Protection Agreement. So far, more than 600 mayors have signed on to its 12-step program for their own cities to meet or beat Kyoto's original target for the U.S.—cutting greenhouse-gas emissions to 7% below 1990 levels by 2012. Some cities got a head start. Portland, Ore., which zeroed in on global warming beginning in 1993, has already slashed emissions 13% per capita, in part by building light-rail lines and 730 miles of regional bikeways—and you can almost smell the clean, green results in the picture of the city's downtown on these pages.

In Austin, Texas, the city-owned utility was able to cancel construction of a 500-MW coal-fired power plant—planned to power 50,000 homes—thanks in part to an intensive green building program that offers energy-efficiency audits to all residents and businesses, retrofits schools and installs insulation and shade screens to reduce sunlight in low-income housing. "We're frustrated by the lack of national leadership," says Mayor Will Wynn, an early backer of the Nickels initiative. "This is about the future of the planet."

Other U.S. cities are crafting their own solutions. St. Paul, Minn., which has had to forgo Winter Carnival ice sculptures and on-ice softball tournaments in recent years because of rising temperatures, is using a biomass-fired power plant for both heat and electricity. Keene, N.H., is harnessing methane and other gases at its landfill to run a generator that powers its recycling center. Salt Lake City, Utah, has converted 1,630 traffic stops to energy-efficient light-emitting diode signals—which alone will avoid the production of more than 500 tons of CO_2 pollution each year and cost the city $53,000 less than conventional bulbs. "The idea is to solve global warming one city at a time," says Glen Brand, an energy specialist for the Sierra Club, which has launched a "cool cities" website.

But though some mayors prefer to downplay the costs of fighting global warming, there seems to be truth to the Bush Administration's contention that meeting the Kyoto targets involves pain—not just gain. Consider Seattle, where population growth is projected to push up regional greenhouse gases 38% in the next 15 years. Ratcheting down to 1990 levels would require slashing the city's emissions by 683,000 tons—the equivalent of taking some 148,000 cars off the road. To do that may require such unpopular measures as highway tolls and increased parking taxes. But in the absence of federal controls, Nickels says, he's ready and willing: "If it's not going to happen from the top down, let's make it happen from the bottom up."

Role model *Portland, Ore., is often called the nation's greenest large city. Here residents promenade along downtown's Tom McCall Waterfront Park, part of the city's extensive network of greenways for biking and hiking*

Visionary *Iceland's President, Olafur Grimsson, has turned his resource-rich nation into a laboratory for researching alternate energy sources*

States are also joining hands to curb emissions from power plants—the coal burned in Pennsylvania, after all, doesn't pause at the New Jersey state line. In 2003 then Governor George Pataki of New York launched the Regional Greenhouse Gas Initiative, a confederation of Northeastern and Mid-Atlantic states that has created its own cap-and-trade program, with the goal of reducing emissions 10% below 2006 levels by 2019. As of July 2007, ten states had joined the group. In February 2007 five Western states embraced a similarly ambitious goal in the Western Regional Climate Action Initiative; a sixth state and two Canadian provinces joined them within months.

Around the world, national governments are taking action. Thanks to President Olafur Grimsson, Iceland may be the stage for the next big advance against global warming. Over the next two years, a team of scientists will try to inject carbon-dioxide-charged water into the basalt beneath Iceland's ground through boreholes drilled by a nearby geothermal energy plant. The CO_2 will, in theory, react with the porous rock and form a stable mineral that could remain in the rock for millions of years. If they're right, Iceland could not only render itself carbon neutral but also give the world a means of protection from the effects of CO_2 emissions until they can be reduced.

Grimsson, 64, has witnessed Iceland's conversion from a coal-dependent economy to a nation that gets most of its heat from clean, renewable geothermal resources in his own lifetime. "If Iceland could achieve such a radical change in one generation, enormous changes can succeed all over the world," he says. Basalt sequestration is one of several efforts to boost Iceland's role in climate-change science, including research into soil carbon sequestration and hydrogen-powered transportation. "We have enormous amounts of clean energy and a small society," notes Sigurdur Gislason, a research professor at the University of Iceland. "You can do experiments here that you can't do anywhere else."

At the commercial level, such surprising players as Wal-Mart and the Midwestern power company Duke Energy are pioneering green business practices. So is the Chicago Climate Exchange, a voluntary but legally binding bourse whose members, according to founder Richard Sandor, account for 8% of the greenhouse emissions from stationary sources in the U.S. "If we were a country," he says, "we'd be roughly the size of Britain." Members of the Chicago exchange, including Ford Motor Co. and DuPont, have pledged to cut their emissions 4% by the end of 2007 from the levels they averaged from 1998 to 2000. They have already taken tens of millions of tons of greenhouse gases out of play—which sounds impressive until it's compared with the 6 billion-ton plume of CO_2 spewed into the atmosphere by the U.S. each year.

Elsewhere, Internet ventures with names like TerraPass, myclimate and Drive Neutral enable travelers to calculate their emissions and neutralize the damage by investing in green practices. Admittedly, such efforts are more about raising environmental consciousness than in making a serious dent in carbon overload. "There is an educational value in these things," says Judi Greenwald of the Pew Center on Global Climate Change. "People realize that what they do can make a difference." From Portland to Iceland, that message seems to be coming across at last, green and clear. ∎

■ The Retailer: Wal-Mart

All around the world, shoppers flock to Wal-Mart to buy everything from socks to sofa beds. In McKinney, Texas, they come for another reason: to see the wind turbine. Rising 120 ft. above the ground, it's the tallest structure in town and supplies 5% of the store's electricity. It's not the only thing that makes this Wal-Mart a green giant. There are photovoltaic shingles on the roof, exterior walls coated with heat-reflective paint and a high-tech system that automatically dims or raises the lights, depending on sun conditions.

Attention, shoppers: Wal-Mart is going green. And the retailer is so big, with 5,200 stores worldwide, that it influences everything from the price of lumber to the size of your cereal box. If Wal-Mart throws its weight behind environmental responsibility, the impact could be profound: less air pollution at factories in China, mass-market sales of organic products, cereal boxes that aren't half filled with air. "One little change in product packaging could save 1,500 trees," says Wal-Mart CEO Lee Scott. "If everybody saves 1,500 trees or 50 barrels of oil, at the end of the day you have made a huge difference."

Scott has promised to cut greenhouse-gas emissions at existing stores 20% over the next few years and pledged to construct new stores that are 25% to 30% more efficient. He wants Wal-Mart's fleet of more than 7,000 trucks to get twice as many miles per gallon by 2015. Factories that show Wal-Mart they're cutting air pollution—even those in China—will get preferential treatment in the supply chain. Wal-Mart says it's also working with consumer-product manufacturers to trim their packaging and will reward those that do so with prime real estate on the shelves.

Cynics might call it "greenwash," a bid to deflect attention from Wal-Mart's controversial labor and health-insurance practices. Scott freely acknowledges that he launched the plan in part to shield Wal-Mart from bad press about its contribution to global warming: "By doing what we're doing today you avoid the headline risks that are going to come for people who did not do anything," he says. "At some point businesses will be held accountable for the actions they take."

Fresh breeze *A wind turbine helps create power for a Wal-Mart store in McKinney, Texas*

■ The Nation: Sweden

Like the U.S., Sweden is addicted to oil. Unlike the U.S., it has a plan to kick the habit—and a deadline. By 2020, says Mona Sahlin, Minister for Sustainable Development, the country will no longer be dependent on fossil fuels. "By then," she declares, "no home will need oil for heating, no motorist will be obliged to use petrol [gasoline] as the sole option available."

Can Sweden do it? Probably. Back in 1970, before the first Middle East energy crisis, Sweden got 77% of its energy from oil. By 2003, even though the nation's industrial production had risen dramatically, that figure had dropped to 34%.

Sweden was rated the world's second greenest nation (just behind New Zealand) in a study issued at the 2006 World Economic Forum in Davos, Switzerland. Its leaders have passed laws that would be unthinkable for a U.S. politician: taxes on fuel and CO_2 emissions to induce car owners to trade in their gas guzzlers for hybrids, and tax exemptions for home owners who switch from oil heating to renewable energy. And Swedes are rushing to convert cars from gasoline to fuels like ethanol and biogas, fuel made from fermented plant waste. Stations that sell alternative fuels are springing up all over the country, and in Helsingborg, a coastal city of 120,000, buses run on biogas made from organic waste from both households and farms.

What Americans might appreciate is the way local governments are encouraged to devise their own strategies for meeting the national goals. The old university city of Lund gets 30% of its heat from a geothermal plant. And Fjaras, in the southwest, just opened a solar-powered health center. Some of these are small efforts, to be sure, but when an entire nation embraces a pledge to wean itself from oil, there's no reason it can't be done.

Beyond coal *Testing a biogas-powered train in Sweden*

■ The Politicians: Michael Bloomberg and Arnold Schwarzenegger

Brighter days *New York City Mayor Michael Bloomberg touts the merits of an energy-efficient light bulb at a 2007 press conference*

On an unseasonably hot May day in Central Park in 2007, New York City Mayor Michael Bloomberg—the pint-size billionaire whose last name needs no elaboration for anyone who knows anything about finance or the media—was talking about saving the planet. With the mayors of more than 30 of the world's largest cities at his side, Bloomberg was opening a climate summit, highlighting his ambitious plan to slash the Big Apple's carbon emissions. Together, the mayors pledged to enlist their 250 million constituents in the fight against global warming. "Unfortunately, partisan politics has immobilized Washington," Bloomberg said. "But the public wants this problem solved. Cities can't wait any longer for national governments to act."

Bloomberg is acting, all right. His PlaNYC calls for the city to cut greenhouse gases 30% by 2030. It seeks to quadruple the city's bike lanes, convert the city's taxis to hybrids and impose a congestion fee for driving into Manhattan (as expected, that proposal met early resistance from New York State legislators).

On Sept. 26, 2006, Arnold Schwarzenegger—the fridge-size multimillionaire whose last name needs no elaboration, period— signed a bill restricting the carbon emissions of factories, utilities and refineries. The former Hollywood action star, who has been Governor of the Golden State since 2003, has already tricked out two of his five Hummers, one to run on biofuel and another on hydrogen. The feds have done nothing on fuel efficiency in two decades, but as of midsummer 2007, 11 states said they would follow California's lead if President George W. Bush grants a waiver excusing states from the federal Clean Air Act, which places emission-regulation authority in Washington's hands.

In May 2007, after signing a climate deal with Ontario, the Terminator said he had a message for Detroit: "Get off your butt!" He had a similar message for Washington. "Eventually, the Federal government is going to get on board," he said. "If not, we're going to sue."

The Hollywood brute and the Wall Street mogul may look like the oddest couple since *Twins,* but there's a reason Arnold calls Bloomberg his soul mate. They're both self-confident, self-made men who rose to stardom from middle-class obscurity. Both have run as Republicans with liberal social values, though Bloomberg made waves in June 2007 when he announced he was leaving the G.O.P. to become an independent, fueling speculation that he might run for the presidency in 2008. Both men are distressed by the lack of environmental leadership at the federal level—and both are thinking globally and acting locally, working hard to make their bailiwicks greener.

"These are two exceptional and forceful guys who don't need the job at all; they had pretty damn good lives before they got into politics," says their mutual friend Warren Buffett. "They're in office to get things done. And they're doing that a lot better than anyone in D.C."

Nowhere is that more evident than on green issues—and Bloomberg and Schwarzenegger aren't acting alone. Washington rejected the Kyoto Protocol, but more than 600 U.S. mayors have pledged to meet its emissions-reduction standards. Denver Mayor John Hickenlooper's Greenprint Denver is an aggressive plan to reduce the city's greenhouse-gas emissions. And Utah Governor Jon Huntsman is the first Republican to join Schwarzenegger and Democrats in signing the Western Regional Climate Action Initiative, a plan that will improve the states' energy efficiency up to 25% and create a regional carbon-trading system. Looks like there's still room for pioneers out West.

"Get off your butt!" *California Governor Arnold Schwarzenegger has become an outspoken advocate for caps on hydrocarbon emissions*

The Utility: Duke Energy

Jim Rogers runs a power company that spews 62 million tons of carbon dioxide into the atmosphere each year. That's a lot of greenhouse gas. But you won't find him on the hit list of environmental crusaders. The CEO of Duke Energy, a utility with nine coal-fired plants in Indiana, Ohio and Kentucky, Rogers is an outspoken advocate of regulating carbon and imposing a price on emissions. His position makes him a renegade within his industry, which officially opposes any regulatory scheme that would force power companies to cut carbon emissions. It makes Rogers more likely to be invited to Sierra Club headquarters than to the White House, given that President George W. Bush has never called for policies more stringent than voluntary cuts in greenhouse gases.

What is Rogers thinking? For one thing, he's personally worried about global warming and believes that the scientific debate about what causes it has long been settled. He thinks that the U.S. will be forced to regulate carbon—as most other industrialized countries have done—within the next few years, if not sooner. And as the CEO of a publicly traded company, he has to make decisions that will affect shareholders decades in the future. Power plants have life spans of 50 years, he notes, and if carbon is taxed, the fuel calculus of those plants changes radically. "We're very dependent on coal," says Rogers, "and if you're going to have earnings growth that's sustainable over a long period of time, you [need] certainty on the carbon issue."

Since the April, 2006 merger of his former company, Cinergy, with Duke Energy, Rogers has run one of America's largest utilities, and he aims to lead by example. In the years before the merger, Cinergy spent $1 billion to increase its use of cleaner-burning natural gas, including $200 million to convert a coal-fired plant, and Rogers cut Cinergy's reliance on coal from 87% of its fuel to 73%. He has pledged to reduce Duke Energy's CO_2 emissions 5% below 2000 levels by 2012, and he is investing in projects to sequester carbon under forests. Rogers is evaluating coal-gasification technology for a plant in Indiana, which could dramatically cut carbon emissions from burning coal, still the least expensive and most abundant fossil fuel in the U.S.

Even if he succeeds, Duke Energy's environmental record will be far from perfect. A $1.4 billion settlement with the Environmental Protection Agency over alleged violations of the Clean Air Act fell apart when Rogers backed away from the deal. The original suit is slowly working its way through the courts. And Duke supports Bush's efforts to roll back provisions of the Clean Air Act that govern utilities.

But with global warming, Rogers vows to keep the heat on his colleagues in the energy industry and on Washington politicians. "My greatest fear is that we don't deal with the problem now," he says, "and we wake up one day and don't have enough time."

■ Pioneers of Green Leadership

Sunita Narain and Bhure Lal

Two determined crusaders help clean up New Delhi's transportation system—and its air

Melting ice caps didn't figure into the fight Sunita Narain and Bhure Lal led to build the world's cleanest public-transport network. They had more pressing concerns. "New Delhi was choking to death," says Narain, 44, director of India's Center for Science and Environment. "Air pollution was taking one life per hour." Adds Lal, 64, then a senior government administrator: "The capital was one of the most polluted on earth. At the end of the day, your collar was black, and you had soot all over your face. Millions had bronchitis and asthma."

In the mid-1990s, Narain filed a lawsuit to force Delhi's buses, taxis and rickshaws to convert to cleaner-burning compressed natural gas (CNG). In July 1998, the Supreme Court ruled largely in her favor and adopted many of her proposals. It ordered a ban on leaded fuel, conversion of all diesel-powered buses to CNG and the scrapping of old diesel taxis and rickshaws. But busmakers, oil companies and some government ministers objected. So the court formed a committee, led by Lal and Narain, to enforce its judgment.

The duo immediately ran into roadblocks: idled buses, angry commuters, long gas-station queues for CNG, oil-company scientists who claimed CNG was as bad as petroleum. But Narain and Lal fought back. By December 2002, the last diesel bus had left Delhi, and 10,000 taxis, 12,000 buses and 80,000 rickshaws were powered by CNG. "Delhi leapfrogged," Narain says with a grin. "People noticed."

Auden Schendler

An indomitable snowman brings a valuable green perspective to skiing's all-white world

If the 1998 fires set in Vail, Colo., by protesters from Earth Liberation Front were an environmental wake-up call for the ski industry, Auden Schendler, 36, is a triple shot of espresso. Hired the next year by Aspen Skiing Co. (ASC), he has become the most visible of a crop of experts charged with cleaning up the industry's act. Between keeping the lodges toasty and draining the creeks for snowmaking, downhill-skiing companies in the late 1990s were major consumers of natural resources. And ASC, which operates four mountains, two hotels and 12 restaurants in the Aspen-Snowmass area, was one of the biggest. Its snowmaking operations alone consume some 160 million gal. of water a year.

Schendler set about changing that. At his urging, ASC became the first ski company to issue a climate-change policy, with a public commitment to cutting greenhouse gases that has led to a 75% reduction in emissions. ASC was the first to use biodiesel fuel in snowcats, issue sustainability reports and develop a green building policy.

A graduate of Bowdoin College, Schendler found a kindred spirit in ASC president and CEO Pat O'Donnell. Even so, it took four years to persuade the company to retrofit a parking garage with fluorescent light fixtures, a move Schendler calculates rid the atmosphere of 300,000 lbs. of CO_2 annually. His goal is to change the culture of skiing, at least at ASC. "We've turned this place into a lab for addressing climate change," Schendler says. "Aspen is a lever that can change the world." Well, Colorado will do for now.

Li Zheng

An engineer helps China chart a path toward cleaning up its emissions from burning coal

Like just about every ambitious engineering student at China's Tsinghua University in the early 1980s, Li Zheng had his heart set on the high-tech, high-profile electronics field—up until the day he bombed on an electronics exam. But his uncharacteristic classroom stumble led Li to a field that could play an even larger role in China's future: energy production. "I think the choice was a very fortunate one in the end," says Li, who studied thermal engineering and in 2000 became a full professor at Tsinghua—China's M.I.T.— at the remarkably young age of 35. "Energy is incredibly important for a growing society like China."

But energy means carbon, and China's booming economy puts it on a path to become the world's No. 1 greenhouse-gas emitter as early as 2020. Li, shown above at a coal-fired power plant, knows that China needs clean energy as badly as the developed world needs China to clean up, which is why he joined the Tsinghua-BP Clean Energy Research and Education Center as director in 2003. The center's most promising project is a new technology called polygeneration, by which coal is converted into a cleaner gaseous fuel that can both generate electricity and be processed into a petroleum substitute. Polygeneration could cut China's carbon emissions and reduce its dependence on oil imports. The technology is still more expensive than direct coal combustion, but Li is lobbying the government to construct a $600 million demonstration plant, and he's optimistic he will see it built.

Peter Liu

A banker with a heart of green helps eco-entrepreneurs fund planet-friendly projects

At the nation's first commercial bank aimed at green businesses, New Resource Bank in San Francisco, founder and vice chairman Peter Liu, 41, finances environmentally savvy, resource-efficient ventures. The institution operates as a full-service communi-

ty bank, but the deposits are used to finance loans for environmentally conscious projects, including alternative energy, clean tech, organic farming and sustainable home and office construction. Green builders get lower interest rates, while home owners can finance solar-power installations for about the same cost as their monthly electricity bill.

Founded in 2006, the bank is already attracting like-minded depositors around the nation, and Liu has plans to expand throughout the U.S. After all, nothing is as green as money.

Greenhouse *Residents of Findhorn Ecovillage in Scotland have been pioneering green practices, such as living in cabins roofed with turf, since the 1980s. The village boasts that it has one of the lowest ecological footprints in the world*

Actions

What should we do? There was a time when "saving the earth," like "saving the whales," was widely used as a catchphrase for futile, grandiose crusades. But global warming challenges us to preserve our planet's climate—or else. On the following pages, we offer hands-on steps each of us can take to mitigate climate change

1 Turn Food into Fuel

ARE CORN HUSKS BETTER THAN CORN FOR PRODUCING ENERGY? ETHANOL IS THE alternative fuel that could finally wean the U.S. from its expensive oil habit and in turn prevent the millions of tons of carbon emissions that go with it. As of 2007, the Department of Energy has doubled its 2005 commitment to funding research into biofuels—any nonpetroleum fuel source, including corn, soybean, switchgrass, municipal waste and (ick) used cooking oil. Already, half of the nearly 11 billion bushels of corn produced each year is turned into ethanol, and most new cars are capable of running on E10 (10% ethanol and 90% gas).

Yet the eco-friendly fuel is beginning to look less chummy of late. Some of the 114 ethanol plants in the U.S. use natural gas and, yes, even coal to run the processors. And ethanol has to be trucked. Existing gas pipelines can't carry it because it corrodes iron. Then there are the economics. Producers depend on federal subsidies, and increasing demand for corn as fuel means the kernels keep getting pricier.

IMPACT	LOW ———▼— HIGH
TIME HORIZON	NOW ———▼— LATER
FEEL-GOOD FACTOR	LOW ▼——— HIGH

That's why researchers are prospecting for more alternatives, preferably ones that don't rely on food crops or a 51¢-per-gal. tax break. Municipal waste, wood pulp and leftover grain and corn husks are all quite attractive; they can produce something called cellulosic ethanol, which contains more energy than corn. But they don't give up their bounty easily, so for now they're more expensive than corn-based ethanol to produce. Undeterred, researchers at several cellulosic-ethanol plants are developing innovative enzyme concoctions and heating methods to make the process more economic. Nothing like haste to make something out of waste.

2 Get Blueprints For a Green House

Reducing your impact on the earth is not just a question of what you drive but also of what you live in. Residential energy use accounts for 16% of greenhouse-gas emissions. If you begin thinking green at the blueprint stage, however, low-tech, pragmatic techniques will maximize your new home's efficiency. Installing green systems from the ground up is cheaper than retrofitting. "Doing simple things could drastically reduce your energy costs, by 40%," says Oru Bose, a sustainable-design architect in Santa Fe, N.M.

For example, control heat, air and moisture leakage by sealing windows and doors. Insulate the garage, attic and basement with natural, nontoxic materials like reclaimed blue jeans. Reduce solar heat in windows with large overhangs and double-pane glass. Emphasize natural cross ventilation. "You don't need to have 24th century solutions to solve 18th century problems," Bose says. Next, consider renewable energy sources like solar electric systems, compact wind turbines and geothermal heat pumps to help power your home. When you're ready to get creative, *GreenHomeGuide.com* will help you find bamboo flooring, cork tiles and countertops made from recycled wastepaper.

IMPACT	LOW ————▼— HIGH
TIME HORIZON	NOW ———▼—— LATER
FEEL-GOOD FACTOR	LOW ————▼ HIGH

3 Change Your Lightbulbs

The hottest thing in household energy savings is the compact fluorescent lightbulb (CFL), a funny-looking swirl that fits into standard sockets. CFLs cost three to five times as much as conventional incandescent bulbs yet use one-quarter the electricity and last several years longer. They are available virtually everywhere lightbulbs are sold.

One caveat: most labels don't say "CFL" (GE calls its bulbs Energy Savers), and in some cases the telltale twist is enclosed in frosted glass. The wattage gives them away: many 7-watt CFLs are comparable to a regular 40-watt bulb, 26 watts is the typical CFL equivalent of 100 watts and so on. Or just look for the Energy Star label.

CFLs have come a long way since they were first introduced in the mid-'90s (they don't flicker as much when you turn them on, for one thing), but because each bulb still contains 5 mg of mercury, you're not supposed to toss them out with the regular trash, where they could end up in a landfill. The bulbs are one more thing for you to sort in the recycling bin.

Light-emitting diodes, or LEDs (see item 4), don't have this problem, but they can require a bit of DIY rewiring. LEDs work great as accents and task lights, and you can also buy LED desk and floor lamps. But if you're just looking to put a green bulb in your favorite table lamp, CFL is the way to go.

IMPACT	LOW ▼——————— HIGH
TIME HORIZON	NOW ▼——————— LATER
FEEL-GOOD FACTOR	LOW ——————▼— HIGH

1 Fuel up with
corn power

4 Light Up Your City

Cities can save energy—and money—by illuminating public spaces with LEDs, or light-emitting diodes. In December 2006, Raleigh, N.C., turned one floor of a municipal parking garage into a testing ground for LEDs (see the before-and-after photos at *cree.com/LEDcity*). The new white, brighter fixtures use 40% less electricity than the high-pressure sodium bulbs they replaced. Although they cost two to three times as much, they can go five or more years without upkeep. Traditional bulbs must be replaced every 18 months. Other types of LEDs are already at work in traffic lights, outdoor displays (like those in New York City's Times Square) and stadiums; airports even use LEDs on their taxiways. If your city is still burning tax money on old lights, ask the mayor why.

IMPACT LOW — HIGH
TIME HORIZON NOW — LATER
FEEL-GOOD FACTOR LOW — HIGH

5 Pay the Carbon Tax

EVERYONE AGREES THAT IT'S NECESSARY TO BEGIN reducing carbon emissions around the world. There is less agreement over exactly how nations should go about achieving a more emissions-free planet. It's the environmental equivalent of Elvis vs. the Beatles: Should nations cap-and-trade carbon emissions or impose a carbon tax on all users? With cap-and-trade programs, governments limit the level of carbon that can be emitted by an industry. Companies that hold their emissions below the cap can sell their remaining allowance on a carbon market, while companies that exceed their limit must purchase credits on that market. Carbon taxes are more straightforward: a set tax rate is placed on the consumption of carbon in any form—fossil-fuel electricity, gasoline—with the idea that raising the price will encourage industries and individuals to consume less.

At the moment, the cap-and-trade system has the upper hand, since it serves as the backbone of the current Kyoto Protocol, and helped the U.S. reduce acid rain in the 1990s—but don't write off the tax just yet. Its supporters argue that a cap-and-trade system, especially one that would be global enough to mitigate the 8 billion tons of carbon the world now emits, would be too difficult to administer—and might be too easily gamed by industries looking to sidestep emissions caps. Cap-and-trade advocates counter that like all other flat taxes, a carbon levy would disproportionately burden lower-income families, who spend a greater percentage of their income on energy than rich households.

So which system will have the largest impact on carbon consumption? A 10% flat carbon tax might reduce the demand for carbon about 5% or less, according to an analysis by the Carbon Tax Center, an environmental advocacy group. That may not be enough. Businesses and governments haven't yet figured out how the two competing regimes can work together, and in the end, the world may need to put both of them to work.

IMPACT LOW — HIGH
TIME HORIZON NOW — LATER
FEEL-GOOD FACTOR LOW — HIGH

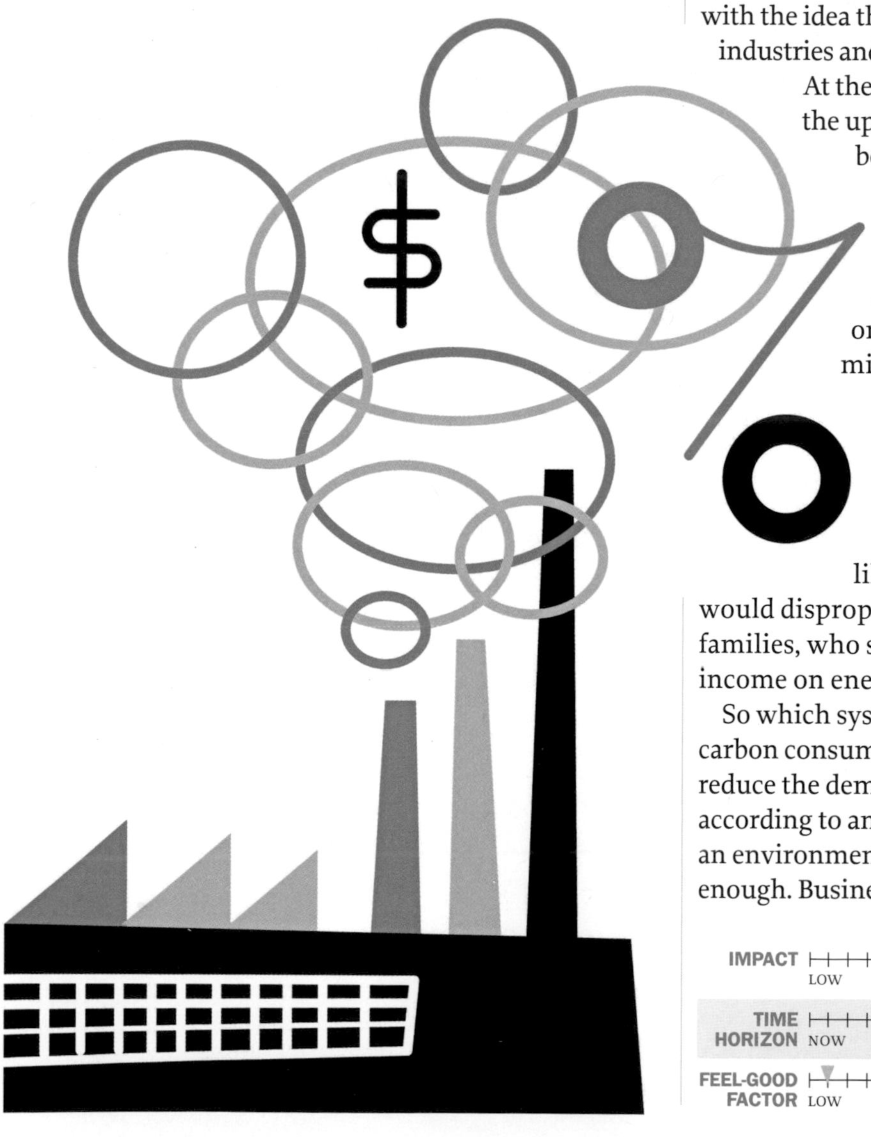

6 Ditch the McMansion

Oversize houses aren't just architecturally offensive; they also generally require more energy to heat and cool than smaller ones, even with efficient appliances. And in the U.S., big houses are becoming the norm, though a relatively inefficient small house consumes less energy than a greener large house. Smaller homes also use fewer building materials, which expand a structure's carbon footprint, than do larger ones.

If you really want to live small, visit Jay Shafer. The former art professor dwells alone in a home fit for a hobbit, 100 sq. ft. in Northern California, that he designed and built himself in 1999. Shafer now runs Tumbleweed Tiny House and sells custom designs for miniature dwellings that range from 70 sq. ft. to 350 sq. ft. He made his move because he felt guilty about the size of his residential carbon footprint, and he now prefers life tiny and tidy. "If I throw my jeans down on the floor, I can't get across the room."

IMPACT	LOW	HIGH
TIME HORIZON	NOW	LATER
FEEL-GOOD FACTOR	LOW	HIGH

7 Hang Up A Clothesline

IMPACT	LOW	HIGH
TIME HORIZON	NOW	LATER
FEEL-GOOD FACTOR	LOW	HIGH

You could make your own clothes with needle and thread using 100% organic cotton sheared from sheep you raised on a Whole Foods diet, but the environmental quality of your wardrobe is ultimately determined by the way you wash it. A recent study by Cambridge University's Institute of Manufacturing found that 60% of the energy associated with a piece of clothing is spent in washing and drying it. Over its lifetime, laundering a single T shirt can send as much as 9 lbs. of carbon dioxide into the air.

The solution is not to avoid doing laundry, tempting as that may be. Rather, wash your clothes in warm water instead of hot, and save up to launder a few big loads instead of many smaller ones. Use the most efficient machine you can find—newer ones can use as little as one-fourth the energy of older machines. Dry your clothes the natural way, by hanging them on a line rather than loading them in a dryer. All told, you can reduce the CO_2 created by your laundry up to 90%. And there's one more bonus: no more magically disappearing socks.

8 Give New Life To Your Old Fleece

Where do old fleece jackets go to die? In some cases, the answer is: back to the mountain. Outdoor-gear label Patagonia is collecting used clothing (regardless of brand) made from Polartec and Capilene to melt and make into new fabric and clothes. (Some of that fleece is especially virtuous; it started out as fabric made from recycled plastic.)

The company estimates that making polyester fiber out of recycled garments, compared with using new polyester, will result in a 76% energy savings and greenhouse gases reduced 71%. To find out more about shearing your own fleece, visit *www.patagonia.com/recycle*.

IMPACT	LOW	HIGH
TIME HORIZON	NOW	LATER
FEEL-GOOD FACTOR	LOW	HIGH

9 Build a Green Skyscraper

IMPACT LOW ———————— HIGH

TIME HORIZON NOW ———————— LATER

FEEL-GOOD FACTOR LOW ———————— HIGH

The greenest building *Five artist renderings of New York City's Bank of America tower*

ALMOST EVERYTHING ABOUT THE BANK OF AMERICA TOWER, A SOARING SKY-scraper under construction in 2007 near Times Square in New York City, has been designed to minimize the use of energy. Take the concrete. Making the stuff from scratch is very energy intensive, so the builders are using a mix of 55% concrete and 45% slag, a waste product from blast furnaces. Mixing slag with concrete saves energy and makes the concrete stronger. The tower will save so-called gray water from washrooms and use it to flush the toilets. The building will also generate much of its own electricity from natural gas, a less potent carbon emitter than coal. These features will account for $3.5 million of a total building cost of $1.2 billion, but the owners expect to recoup that in a few years with all the energy they will save. When it's finished in 2008, the tower will be the second highest in the city, but it will stand alone as the greenest building in New York.

10 Use Geothermal Heating

With clever engineering and an elegantly simple design, Diane von Furstenberg reinvigorated women's fashion in the 1970s with the wrap dress. Can she do the same for a building? Her newest project is a 35,000-sq.-ft. office, showroom and retail store in Manhattan's trendy Meatpacking District, all heated and cooled by water pumped from deep underground. The building's geothermal system taps into water that is a relatively stable 55°F and transfers that heat to warm the building in the winter and cool it in the summer.

The building's roof is home to easy-to-maintain plants and grasses; it also has two heliostat mirrors, which track the sun and direct its rays into the building, reducing the use

of artificial lights during the day. Who says being environmentally conscious can't be cool—and hot?

IMPACT LOW ———————— HIGH

TIME HORIZON NOW ———————— LATER

FEEL-GOOD FACTOR LOW ———————— HIGH

11 Take Another Look At Vintage Clothes

High-end hand-me-downs (the smart set calls them vintage) are more ecologically sound than new clothes. Why? Buying a shirt the second time around means you avoid consuming all the energy used in producing and shipping a new one and, therefore, the carbon emissions associated with the transaction.

Every item of clothing you own has an impact on the environment. Some synthetic textiles are made with petroleum products. Cotton accounts for less than 3% of farmed land globally but consumes about a quarter of the pesticides. One quick way to change your duds: invite friends over for a closet swap, to which everyone brings a few items they want to trade. It's easy on both the environment—and your pocketbook.

IMPACT LOW ———————— HIGH

TIME HORIZON NOW ———————— LATER

FEEL-GOOD FACTOR LOW ———————— HIGH

12 Capture The Carbon

Coal is one of the dirtiest fuels around and a major source of the planet's carbon dioxide emissions. It's also hard to live without: one-half the electricity generated in the U.S. comes from coal. So here's a thought: What if coal-fired plants stopped spewing their carbon dioxide fumes into the air and instead sequestered them—pumped them deep into the ground for storage?

Carbon sequestration is (despite its name) a simple-sounding idea that's exciting scientists, governments and energy companies as a way to cut emissions without disrupting energy supplies. One coal-fired plant in Denmark is working to trap carbon flue gases and store them in four spots, including an unused oil field off the coast of Spain. A Swedish utility is testing new ways to extract pure carbon dioxide from coal emissions at a lignite plant in eastern Germany.

In the biggest test of sequestration so far, a Norwegian energy firm is injecting 1 million tons of CO_2 a year from the Sleipner gas field into a saline aquifer under the North Sea. "All the basic technology is already here," says Howard Herzog, an energy expert at the Massachusetts Institute of Technology. A report by the International Energy Agency (IEA) in Paris says sequestration would be second only to energy-saving measures in reducing CO_2 emissions, far ahead of better-known efforts like renewable energy.

There are two major obstacles. The first is cost, which the IEA estimates to be as much as $50 for each ton of carbon captured. Those costs may drop if the technology is successful and utilities are given incentives not to spew out carbon dioxide. The other obstacle is a lack of detailed scientific knowledge. The pilot projects are going well, but, M.I.T.'s Herzog says, "we'd like to see more large-scale demonstrations worldwide to really bolster confidence." In the meantime, watch for sequestration to move quickly up the world's energy to-do list.

IMPACT	LOW — HIGH
TIME HORIZON	NOW — LATER
FEEL-GOOD FACTOR	LOW — HIGH

121

13 Let Employees Work Close to Home

SITTING IN GRIDLOCK WASTES YOUR TIME AND THE planet's fuel. The only solution, it seems, is to move your home next to the office. But what if you could move the office a little closer to home?

That, in essence, is the concept called proximate commuting. It works best for companies with multiple locations in one metro area.

Gene Mullins, a software developer in Seattle, created a program that helps firms slash the time employees spend driving by matching them with work closer to home. Mullins did studies for Starbucks, Key Bank, Boeing and, most recently, Seattle's fire department. He found that only 4% of the firefighters worked at the station closest to their home; some commuted 145 miles each way. At Boeing, daily commutes of its 80,000 Puget Sound employees total 85 circumnavigations of the earth. Using Mullins' program, a number of Key Bank branches reduced commutes of some workers 69%. Still, only about 20% of its employees work at the branch closest to their home, Mullins says. Yet escaping rush-hour traffic is its own reward. "For the same pay and the same job, who wouldn't want a shorter commute?"

14 Ride The Bus

Transportation accounts for more than 30% of U.S. carbon dioxide emissions. So one of the best ways to reduce them is by riding something many of us haven't tried since the ninth grade: a bus. Public transit saves an estimated 1.4 billion gal. of gas annually, which translates into about 1.5 million tons of CO_2, says the American Public Transportation Association.

Unfortunately, 88% of all trips in the U.S. are by car. In part, that's because public transportation is more readily available in big urban areas. One promising alternative is bus rapid transit (BRT), which features extra-long carriers running in dedicated lanes. Buses emit more carbon than trains, but that can be minimized by using hybrid or compressed-natural-gas engines. A 2006 study by the Breakthrough Technologies Institute found that a BRT system in a medium-size U.S. city could cut emissions by as much as 654,000 tons over 20 years.

Owing to high gas prices, miles driven per motorist dropped in 2005 for the first time since 1980, according to the Pew Research Center. The U.S. is ready to change. We're just waiting for the bus.

15 Move to A High-Rise

If you're a true Environmentalist, a dyed-in-the-wool greenie, then why not pack up your leafy rural home and move to New York City—preferably to a tall building right in the middle of Manhattan?

The Big Apple is home to the greenest citizens in the U.S. Relatively few New Yorkers own cars—usually the biggest source of an individual's carbon emissions. Most walk, bike or ride public transit to work—all more efficient transport than the best hybrids.

And New York has developed up, rather than out, which limits wasteful sprawl. Eight million New Yorkers are squeezed into 301 sq. mi.—less than a fortieth of an acre per person. Even a fairly dense suburb devotes about a third of an acre to each person. Density means that commutes, shopping trips and supply chains are shorter. Plus, New Yorkers tend to live in small spaces, although they're a little cranky about it. The denser the area you call home, the smaller your personal carbon footprint—and your gas and electricity bill.

16 Pay Your Bills Online

Eliminating your paper trail by banking and paying bills online does more than save trees. It also helps reduce fuel consumption by the trucks and planes that transport paper checks. If every U.S. home viewed and paid its bills online, the switch would cut solid waste by 1.6 billion tons a year and curb greenhouse-gas emissions by 2.1 million tons a year, according to Javelin Strategy & Research.

Worried about security? Don't be. Just ignore e-mails "phishing" for personal data, and monitor all (electronic) statements for any unauthorized debits. Report problems immediately, and your credit won't take the hit.

To avoid unnecessary carbon dioxide–emitting car trips to the bank on payday, ask your employer to directly deposit your paycheck. You will get your money faster that way too.

IMPACT	LOW ——————— HIGH
TIME HORIZON	NOW ——————— LATER
FEEL-GOOD FACTOR	LOW ——————— HIGH

17 Open a Window

MOST OF THE 25 TONS OF CO_2 EMISSIONS EACH AMERICAN IS RESPONSIBLE FOR each year come from the home. Here are some easy ways to get that number down in a hurry without rebuilding. Open a window instead of running the AC. Adjust the thermostat a couple of degrees higher in summer and lower in winter. Caulk and weatherstrip all your doors and windows. Insulate your walls and ceilings. Use the dishwasher only when it's full. Install low-flow showerheads. Wash your clothes in warm or cold water. Turn down the temperature on the water heater. At the end of the year, don't be surprised if your house feels lighter. It just lost 4,000 lbs. of carbon dioxide.

IMPACT	LOW ——————— HIGH
TIME HORIZON	NOW ——————— LATER
FEEL-GOOD FACTOR	LOW ——————— HIGH

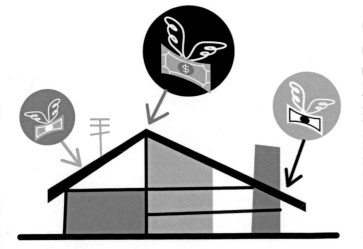

18 Ask the Experts for an Energy Audit of Your Home

How green is your abode? A home energy audit, which most utility providers will do free of charge, will tell you the amount of power your household consumes and what you can do to reduce it. The average family can find ways to shave 1,000 lbs. of CO_2 emissions each year. Energy auditors use special equipment like blower doors and infrared cameras to help you pinpoint exactly how and where your house is losing energy. You can also try a do-it-yourself audit, but this is one time you might actually want to be audited by the experts.

IMPACT	LOW ——————— HIGH
TIME HORIZON	NOW ——————— LATER
FEEL-GOOD FACTOR	LOW ——————— HIGH

19 Buy Green Power, At Home or Away

MORE THAN 600 UTILITIES IN 37 STATES OFFER GREEN ENERGY, BUT UNLESS YOU read the fine print on your bill, you may not know if your power company is one of them. (To find out, visit *eere.energy.gov/greenpower.*) If you don't live in a green power zone, you can support the industry by buying renewable energy certificates, which allow you to purchase green energy in another part of the country. The extra dollars will dispense green power to the national power grid.

IMPACT LOW — HIGH

TIME HORIZON NOW — LATER

FEEL-GOOD FACTOR LOW — HIGH

A wind turbine in the Netherlands *Renewable energy certificates support the green power industry in the U.S.*

20 Check The Label

You wouldn't buy a car without knowing its gas mileage. Why not do the same when choosing energy-efficient ovens or even supermarkets and hotels? Energy Star, a rating system sponsored by the Environmental Protection Agency, will help you find them. Approved products can be pricier, but they cost less to power.

Commercial buildings account for nearly 18% of U.S. greenhouse-gas emissions, but those with the Energy Star label consume 35% less energy than the average. By using Energy Star appliances at home, consumers can reduce their utility bill as much as 30%.

energy ★
ENERGY STAR

IMPACT LOW — HIGH

TIME HORIZON NOW — LATER

FEEL-GOOD FACTOR LOW — HIGH

IMPACT LOW — HIGH

TIME HORIZON NOW — LATER

FEEL-GOOD FACTOR LOW — HIGH

21 Cozy Up to Your Water Heater

Improving your home's efficiency doesn't have to mean hours in the attic tearing out and replacing the insulation. It might be as simple as giving your dear old water heater a warm hug. Wrapping your heater in an insulated blanket—one costs about $10 to $20 at home centers—could save your household about 250 lbs. in CO_2 emissions annually.

Most water heaters more than five years old are constantly losing heat and wasting energy because they lack internal insulation. If the surface feels warm to the touch, get your heater an extra blankie. You will both feel better.

22 Skip the Steak

Which is responsible for more global warming: your BMW or your Big Mac? Believe it or not, it's the burger. The international meat industry generates roughly 18% of the world's greenhouse-gas emissions—even more than transportation—a 2006 report from the U.N.'s Food and Agriculture Organization determined.

Much of that comes from the nitrous oxide in manure and the methane that is, as the New York *Times* delicately put it, "the natural result of bovine digestion." Methane has a warming effect that is 23 times as great as that of carbon dioxide, while nitrous oxide is 296 times as great.

There are 1.5 billion cattle and buffalo on the planet, along with 1.7 billion sheep and goats. Their populations are rising fast, especially in the developing world. Global meat production is expected to double between 2001 and 2050. Given the amount of energy consumed raising, shipping and selling livestock, a 16-oz. T-bone is like a Hummer on a plate. Switching to vegetarianism can shrink your carbon footprint by up to 1.5 tons of carbon dioxide a year, according to research by the University of Chicago.

IMPACT ├──────▼──────┤
LOW   HIGH

TIME ▼────────────┤
HORIZON NOW   LATER

FEEL-GOOD ├──────────▼──┤
FACTOR LOW   HIGH

California Governor Arnold Schwarzenegger may have signed the world's toughest anti-global-warming law, but it is Democrat Terry Tamminen, his environmental adviser, who is emerging as the state's real Terminator, winning industry support and the endorsement of his Republican boss for a mandate to reduce the state's emissions 80% by 2050.

Thwarting climate change isn't a solo effort. Tamminen left his official post to build a national response to global warming one state at a time. "I am trying to Johnny Appleseed what California has done," he says. His goal is to create a de facto national climate plan out of individual efforts in the 50 states. "He is crisscrossing the country and spreading the word," says Karl Hausker, deputy director for the Center for Climate Strategies, a nonpartisan, nonproit group that handles the technical details after Tamminen plants his seeds. Nineteen states have developed or are developing aggressive climate plans. Says Tamminen: "By the time that there is a new Administration in the White House, a majority of Americans will live in states with a meaningful plan that deals with the climate-change issue."

IMPACT ├────────────▼┤
LOW   HIGH

TIME ├──────────▼──┤
HORIZON NOW   LATER

FEEL-GOOD ├─▼──────────┤
FACTOR LOW   HIGH

24 Just Say No To Plastic Bags

ATTENTION, SHOPPERS! THOSE WHITE PLASTIC BAGS YOU bring home from the supermarket and then discard don't self-destruct; they probably end up in a landfill. Every year, more than 500 billion plastic bags are distributed, and less than 3% of those bags are recycled. They are typically made of polyethylene and can take up to 1,000 years to biodegrade in landfills that emit harmful greenhouse gases. Reducing your contribution to plastic-bag pollution is as simple as using a cloth bag (or one made of biodegradable plant-based materials) instead of wasting plastic ones. For your next trip to the grocery store, BYOB.

IMPACT ├──────────▼──┤
LOW   HIGH

TIME ▼────────────┤
HORIZON NOW   LATER

FEEL-GOOD ├──────────▼──┤
FACTOR LOW   HIGH

Climate czar *Terry Tamminen*

25
Support Your Local Farmer

Fruit, vegetables, meat and milk produced closer to home rack up fewer "petroleum miles" than foods trucked cross-country to your table. How do you find them? Search *local-harvest.org* by ZIP code for farmers' markets, greengrocers and food co-ops in your area. The website, which includes handy contact information in its directory listings, also identifies restaurants that specialize in regional and seasonal ingredients.

Want to get closer to the farm? Join a Community Supported Agriculture project, which lets you buy shares in a farmer's annual harvest. In return, you will get a box of fresh produce every week for a season. It will take more than a few visits to the farm stand to reduce the carbon impact of the U.S. food supply. In the meantime, here's another reason to go local: the taste is great.

IMPACT ⊢┼┼┼┼▽┼┼┼┼┤
LOW   HIGH

TIME ▽┼┼┼┼┼┼┼┼┤
HORIZON NOW   LATER

FEEL-GOOD ⊢┼┼┼┼┼┼┼▽┼┤
FACTOR LOW   HIGH

26
Plant Bamboo

Bamboo makes a beautiful fence, and because it grows so quickly (as much as 1 ft. a day or more, depending on the species), it absorbs more CO_2 than, say, a rosebush. Most homeowners have to restrict its growth, lest it get out of control. Do this, however, and you reduce bamboo's capacity as a carbon sink. Only large-scale plantings, which absorb CO_2 faster than they release it, can favorably tip the scales.

IMPACT ⊢▽┼┼┼┼┼┼┼┤
LOW   HIGH

TIME ⊢▽┼┼┼┼┼┼┼┤
HORIZON NOW   LATER

FEEL-GOOD ⊢┼┼┼┼┼┼┼▽┼┤
FACTOR LOW   HIGH

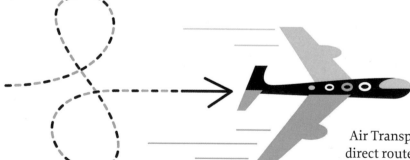

IMPACT ⊢▽┼┼┼┼┼┼┼┤
LOW   HIGH

TIME ⊢▽┼┼┼┼┼┼┼┤
HORIZON NOW   LATER

FEEL-GOOD ⊢┼▽┼┼┼┼┼┼┤
FACTOR LOW   HIGH

27   Straighten Up and Fly Right

UNTIL THE DAY WE CAN TRAVEL BY FIREPLACE, LIKE Harry Potter, the only way to get from Los Angeles to London is by carbon-spewing jet airliner. One simple change can help: adjust the exit and entry points each nation sets for its airspace so that planes can fly in as straight a line as possible. In 2006 the International Air Transport Association negotiated a more direct route from China to Europe that shaved an average 30 minutes off flight time, eliminating 84,800 metric tons of CO_2 annually. Unifying European airspace as a "single sky" could cut fuel use up to 12%.

Pilots could also change the way they fly. Abrupt drops in altitude waste fuel, so experts are advocating "continuous descent" until the plane reaches the runway—where it could be towed instead of burning fuel while taxiing. Of course, the best way to reduce plane emissions is to fly less. At least until the fireplace is ready for takeoff.

28 Tie the Knot with A Green Wedding

YOU WON'T BE ABLE TO STOP GLOBAL WARMING ON YOUR WEDDING DAY, BUT your choices can lessen the carbon footprint of your event. If your guests are traveling long distances, offset the carbon emissions from their trips with a donation to renewable-energy projects. Wherever you celebrate, you can reduce your CO_2 impact and often save money by giving your wedding a local touch. Buy wine from a nearby vineyard or beer from a neighborhood brewery. Get your wedding cake from a local bakery, and use seasonal flowers, not imports. Consider it your wedding gift to the planet.

IMPACT								
LOW								HIGH

TIME HORIZON								
NOW								LATER

FEEL-GOOD FACTOR								
LOW								HIGH

30
Shut Off the Computer

A screen saver is not an energy saver. According to the U.S. Department of Energy, 75% of all the electricity consumed in the home is standby power used to keep electronics running when all those TVs, DVRs, computers, monitors and stereos are "off." The average desktop computer, not including the monitor, consumes from 60 to 250 watts a day.

Compared with a machine left on 24/7, a computer that is in use four hours a day and turned off the rest of the time would save you about $70 a year. The carbon impact would be even greater. Shutting it off would reduce the machine's CO_2 emissions 83%, to just 63 kg a year.

IMPACT	
	LOW — HIGH
TIME HORIZON	NOW — LATER
FEEL-GOOD FACTOR	LOW — HIGH

31
Wear Green Eye Shadow

Bright green may not be in vogue this year, but eco-friendly makeup has trend written all over it. In February '07, Cargo Cosmetics launched PlantLove, a botanical lipstick packaged in a 100% bio-

degradable tube made of poly-lactic acid—a corn-based renewable resource. When the tube is empty, plant it in the ground, and it sprouts flowers. The product represents only a sliver of the $50 billion industry in the U.S., but the trend is growing: the market for organic personal-care products will increase more than 8% in 2007.

IMPACT	
	LOW — HIGH
TIME HORIZON	NOW — LATER
FEEL-GOOD FACTOR	LOW — HIGH

32 Kill the Lights At Quitting Time

Assigning an office switch-off monitor might sound a little like third grade, but it could cut carbon emissions by reducing electricity use, not to mention extending equipment life and lowering maintenance costs. It's not exactly glamorous work: walking the halls to make sure that computers, monitors, desk lights, printers and fax machines are turned off daily.

Air conditioners and overhead lights can be timed for turnoff: Aim for off-peak energy use to be about one-fifth of peak use. In the morning, the switch-on monitor takes over.

IMPACT	
	LOW — HIGH
TIME HORIZON	NOW — LATER
FEEL-GOOD FACTOR	LOW — HIGH

IMPACT	
	LOW — HIGH
TIME HORIZON	NOW — LATER
FEEL-GOOD FACTOR	LOW — HIGH

29 Remove the Tie

HOW CAN A TIE HELP FIGHT CLIMATE CHANGE? When you leave it at home. In the "cool biz" summer of 2005, Japanese salarymen swapped their trademark dark-blue business suits for open collars and light tropical colors. It was all part of the Japanese government's effort to save energy by keeping office temperatures at 82.4°F throughout the summer.

The policy caused sartorial confusion but did make a dent in Japan's rising carbon emissions. In one summer, Japan cut an estimated 79,000 tons of CO_2. If U.S. businesses eased off on their arctic-level air-conditioning, the gains could be significant. Time to make every summer day casual Friday?

33 Rearrange the Heavens and the Earth

WHAT IF WE COULD BUILD A GIANT MIRROR IN SPACE TO DEFLECT THE SUN'S energy? Or inject sulfur into the stratosphere to cool the earth? Scientists are examining such sci-fi methods as a gigantic Plan B should efforts to end man-made carbon emissions fail. Geoengineering, as the developing field is called, involves rearranging the environment on a planetary scale. That these far-out strategies are getting a serious hearing in mainstream science is a measure of how desperate the battle against climate change is becoming.

IMPACT LOW HIGH

TIME HORIZON NOW LATER

FEEL-GOOD FACTOR LOW HIGH

34 Rake in the Fall Colors

Few things rip through the serenity of a Sunday in suburbia like the 70-dB wail of a gas-powered leaf blower. Recent improvements have made the machines more efficient, but using that motorized hurricane for just an hour still sucks up 1 pt. of gas and oil. With more than 30 million acres of lawn in the U.S., it's a high price to pay for a job that can be done almost as well, if somewhat more slowly, with a rake. Besides, you can't lean on a leaf blower when you're done.

IMPACT LOW HIGH

TIME HORIZON NOW LATER

FEEL-GOOD FACTOR LOW HIGH

35 End the Paper Chase

Americans recycled 42 million tons of paper last year—50% of what they used—but still threw out the rest. Paper does grow on trees: 900 million of them every year become pulp and paper. We can reduce that number by buying more recycled paper. It uses 60% less energy than virgin paper. Each ton purchased saves 4,000 kW-h of energy, 7,000 gal. of water and 17 trees, and a tree has the capacity to filter up to 60 lbs. of pollutants from the air.

IMPACT LOW HIGH

TIME HORIZON NOW LATER

FEEL-GOOD FACTOR LOW HIGH

36 Play the Market

IMPACT LOW HIGH

TIME HORIZON NOW LATER

FEEL-GOOD FACTOR LOW HIGH

To cut back on carbon, environmentalists are using the force of the free market. In carbon-emissions trading, the government puts a cap on how much carbon an industry is allowed to emit from power plants, factories and cars. Innovative firms could meet those caps through actual reductions and earn carbon "credits," which they could sell to industry laggards. New York, Connecticut, Delaware, Maine, New Hampshire, New Jersey and Vermont have agreed on a regional cap-and-trade system. Arizona, California, New Mexico, Oregon and Washington have signed a similar pact. New emissions-reduction technology is sexier, but old-fashioned horse trading might just be more effective.

37 Think Outside The Packaging

PAPER OR PLASTIC? WELL, HOW ABOUT neither? All those Styrofoam peanuts and impregnable plastic CD cases cost energy to manufacture and deliver, and that means carbon. You can reduce the amount of packaging you waste with a little consumer vigilance. Give back the extra napkins or unwanted sugar packets; carry that gallon of milk by its handle. True eco-nerds will even bring their own cup to a Starbucks. Corporations are also beginning to pitch in. Hewlett-Packard announced in 2007 that it would switch to lighter packaging for its printer cartridges, which will reduce carbon emissions by an amount equivalent to removing 3,500 cars from the road for a year.

IMPACT: LOW — HIGH
TIME HORIZON: NOW — LATER
FEEL-GOOD FACTOR: LOW — HIGH

Megaretailer Wal-Mart is far out front of most U.S. companies in moving to green packaging. The company has trimmed everything from its rotisserie-chicken boxes to its water bottles, now made with 5 g less plastic. The company plans to cut packaging 5% starting in 2008—enough to prevent 667,000 tons of carbon dioxide emissions.

38 Trade Carbon for Capital

IMPACT: LOW — HIGH
TIME HORIZON: NOW — LATER
FEEL-GOOD FACTOR: LOW — HIGH

Thee Kyoto Protocol's Clean Development Mechanism (CDM) allows companies in wealthier nations to earn emissions-reduction credits for investing in energy-efficient projects in poorer nations. The idea has created a global trade in carbon credits in addition to the broader emissions-trading market and funded hundreds of projects that save about 115 million tons of CO_2 a year.

There have been some hiccups. A recent study found that factories in China were using relatively cheap cleaning systems and then exploiting a loophole to claim hundreds of millions of dollars in carbon credits. But that's no reason to abandon the CDM, argues Rajesh K. Sethi, secretary of India's CDM Authority in the Ministry of Environment and Forests. The CDM is "one of the most successful ways we've found to reduce greenhouse gases," says Sethi. "It needs to be extended, not abandoned."

39
Make Your Garden Grow

The U.S. spends more than $5 billion a year on fossil-fuel-derived fertilizers that leak harmful chemicals into the ground and speed up the release of nitrous oxide, a greenhouse gas. Try alternatives, from trusty compost to grass clippings, which are about 4% nitrogen. Adventurous gardeners use a homemade fertilizer mix that includes seaweed extracts for potassium and fish proteins and oils for nitrogen. Or go native and embrace wildflowers and indigenous grasses. Weeds are a matter of taste.

IMPACT LOW — HIGH

TIME HORIZON NOW — LATER

FEEL-GOOD FACTOR LOW — HIGH

40
Get a Personal Carbon Budget

IMPACT LOW — HIGH

TIME HORIZON NOW — LATER

FEEL-GOOD FACTOR LOW — HIGH

The essential injustice of global warming is that the world's poor will suffer the worst effects of climate change while contributing far less to carbon emissions than the rich. So here's a radical solution: divide greenhouse-gas emissions by population, and give everyone in the world the right to emit the same amount of carbon— a personal carbon allowance.

Essentially, this sort of allowance is a cap-and-trade scheme for individuals. It sets a clear target and lets the market work out the details. Bike to work and live beneath your allowance, and you can sell your carbon credits to energy spendthrifts who refuse to give up their SUVs. The balance of one's allowance might be recorded on a sort of carbon-debit card, so those who do buy that SUV will be spending carbon too. That way, people who want to keep living as if it's 1989 will have to pay for their fantasy.

41 Fill 'er Up with Passengers

THE NEXT TIME YOU GET BEHIND THE WHEEL OF your car, turn to the passenger seat. Chances are, it's empty. In most of the U.S., the single-occupant driver still reigns supreme. Nearly 80% of people drive to work alone; about 38% drive alone in general. In some places, that's starting to change. As part of its Clean Air Act, Washington State appealed to business with incentives to encourage employees to drive less or at least stop driving alone. A state tax credit benefits companies that encourage their employees to carpool, ride the bus, walk or bike to their job, or work fewer days a week. The result: about 20,000 fewer vehicle trips each morning since the program started in 1991, saving commuters $13.7 million and 5.8 million gal. of gas and preventing the release of 78,000 tons of air pollutants and CO_2-equivalent gases.

IMPACT LOW — HIGH

TIME HORIZON NOW — LATER

FEEL-GOOD FACTOR LOW — HIGH

42
Pay for Your Carbon Sins

Feeling full of climate-change guilt, Americans are snapping up carbon offsets from Web-based retailers and nonprofits. Unlike mandatory allowances, offsets allow consumers to pay voluntarily to reduce carbon emissions by a quantity equal to their estimated contribution. The payments typically fund clean-energy projects, pollution control efforts, tree planting and forest conservation.

But offsets are picking up skeptics along with customers. Critics say consumers have little assurance that the projects they underwrite really reduce emissions and warn that those buying offsets may sometimes be paying for improvements that would have happened anyway. They also argue that carbon-offset trading distracts from the urgent need to change U.S. policies to address global warming.

Are these criticisms fair? "There needs to be more standardization, more verification and more assurances for the consumer that the offsets are real," concedes Ricardo Bayon, director of Ecosystem Marketplace. Some organizations, including the Center for Resource Solutions in San Francisco and the Climate Group, based in Britain, are racing to establish certification standards. Even supporters

IMPACT LOW — HIGH

TIME HORIZON NOW — LATER

FEEL-GOOD FACTOR LOW — HIGH

of offset trading agree that it's no substitute for comprehensive national policies. "This voluntary stuff is an interim measure," says Judi Greenwald of the Pew Center on Global Climate Change. "But it is certainly better than doing nothing."

A foggy day—not!
*An artist rendering
of the new green district
in Britain's capital*

43 Move to London's New Green Zone

Homes in London account for 44% of the city's CO_2 emissions. That's why, in the Docklands, builders plan in 2010 to open the city's first large-scale zero-carbon housing development. All 233 homes will hook up to a combined heat-and-power plant that turns wood chips into electricity and hot water, with extra juice from solar panels and wind. Car and bike clubs will lower commuting emissions. The whole project could cost just 5% more than conventional housing developments.

IMPACT LOW — HIGH

TIME HORIZON NOW — LATER

FEEL-GOOD FACTOR LOW — HIGH

44 Check Your Tires

So you own a plain-vanilla, nonhybrid, American-made gas guzzler and can't afford (or can't wait for) a hybrid. Now what? Just giving your engine a tune-up can improve gas mileage 4% and often much more. Replacing a clogged air filter can boost efficiency 10%. And keeping tires properly inflated can improve gas mileage more than 3%. The bottom line? If you can boost your gas mileage from 20 to 24 m.p.g., your old heap will put 200 fewer pounds of CO_2 into the atmosphere each year.

IMPACT LOW — HIGH

TIME HORIZON NOW — LATER

FEEL-GOOD FACTOR LOW — HIGH

45 Make One Right Turn after Another

United Parcel Service took a detour to the right on its way to curb CO_2 emissions. In 2004, UPS announced that its drivers would avoid making left turns. Reason: the time spent idling while waiting to turn against oncoming traffic burns fuel and costs millions each year. A software program maps a customized route for every driver to minimize lefts. In metro New York City, UPS reduced CO_2 emissions by 1,000 metric tons in the first six months of 2007. Today 83% of UPS facilities are heading in the right direction; within two years, the policy will be adopted nationwide.

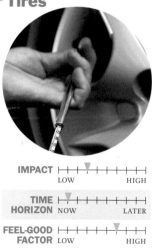

IMPACT LOW — HIGH

TIME HORIZON NOW — LATER

FEEL-GOOD FACTOR LOW — HIGH

Picture of destruction
Hurricanes damage forests. After one tore through southern Sweden in 2005, cleared logs leave the image of a tree

46 Plant a Tree in the Tropics

IT SEEMS LIKE SIMPLE ARITHMETIC: A TREE CAN ABSORB UP TO A TON OF carbon dioxide over its lifetime, so planting one should be an easy way to mitigate climate change—and planting many of them should make a real difference. Turns out it's not so simple: where you plant the trees is what matters. Recent studies have shown that trees in temperate latitudes— including most of the U.S.—actually have a net warming effect on the climate. Why? The heat absorbed by dark leaves outweighs the carbon they soak up.

IMPACT — LOW ——— HIGH

TIME HORIZON — NOW ——— LATER

FEEL-GOOD FACTOR — LOW ——— HIGH

47 If You Burn Coal, Do It Right

The poor, much maligned coal plant: our current versions of this workhorse of energy production not only emit compounds that damage the environment, but they also are not even efficient. More than half the heat the average coal plant generates is simply lost when coal is burned—yes, it goes up in smoke. But in co-generation power plants, that excess heat is captured and reused for domestic and industrial heating, nearly doubling a plant's efficiency.

No one believes coal plants, no matter how efficient we make

IMPACT — LOW ——— HIGH

TIME HORIZON — NOW ——— LATER

FEEL-GOOD FACTOR — LOW ——— HIGH

them, are the ideal solution to the world's growing energy problems, but like it or not, thermal power will remain the backbone of our electricity grid for the foreseeable future. If we're going to continue to burn coal and oil, we might as well make sure all that carbon doesn't go to waste.

48 Drive Green on the Scenic Route

Going on vacation doesn't have to mean leaving your green conscience at home. Car-sharing service Zipcar rents hybrid cars in five U.S. cities, Toronto and London. A few specialty companies offer rental cars that run on biodiesel fuel, a clean-burning substance derived from renewable sources like vegetable oil. Bio-Beetle rents eco-friendly cars, ranging from Passats to Jeeps, in Hawaii and Los Angeles. A week's rental in L.A. runs from $200 to $300. And competitor EV Rental

IMPACT — LOW ——— HIGH

TIME HORIZON — NOW ——— LATER

FEEL-GOOD FACTOR — LOW ——— HIGH

Cars has started to expand beyond the West Coast.

133

49 Set a Higher Standard

If cars have to meet energy standards, why don't power plants? A few states have standards limiting the amount of CO_2 that a new power plant can spew. California's tough 2007 rules virtually exclude new coal plants until clean-coal technology comes online, and could establish a national standard—just as they might for auto emissions. A federal carbon standard would be aggressively opposed by power companies that depend on coal. But it could also spur investment in renewables, clean coal and even nuclear power more rapidly than carbon taxes or cap-and-trade systems. With 159 new coal-powered plants slated for the next decade, a critical choice is looming.

IMPACT   LOW — HIGH
TIME HORIZON   NOW — LATER
FEEL-GOOD FACTOR   LOW — HIGH

50 Be Aggressive About Passive

Georg Zielke, his wife and kids share a five-bedroom "passive house" in Darmstadt, Germany, with heating costs 90% lower than their neighbors'. Extra insulation and state-of-the-art ventilation recycle the energy from passive sources such as body heat, the sun and household appliances to warm the air. A passive house costs 5% to 8% more to build than a standard one. But when it gets really cold, the Zielkes just turn on the TV.

IMPACT   LOW — HIGH
TIME HORIZON   NOW — LATER
FEEL-GOOD FACTOR   LOW — HIGH

51 Consume Less, Share More, Live Simply

THE CHANCE TO BUY A CARBON OFFSET—IN ESSENCE, A SORT OF EMISSIONS indulgence—appeals to the environmental sinner in all of us. But there is an older path to reducing our impact on the planet that will feel familiar to Evangelical Christians and Buddhists alike. Live simply. Meditate. Consume less. Think more. Get to know your neighbors. Borrow when you need to and lend when asked. E.F. Schumacher praised that philosophy this way in *Small Is Beautiful:* "Amazingly small means leading to extraordinarily satisfying results."

IMPACT   LOW — HIGH
TIME HORIZON   NOW — LATER
FEEL-GOOD FACTOR   LOW — HIGH

We do not inherit the earth from
our ancestors, we borrow it from
our children.
        —Native American proverb

Resources

Forecast: moist *This temperate rain forest in Olympic National Park in Washington State is the wettest single area in the continental U.S.*

BOOKS

Greenhouse: The 200-Year Story of Global Warming
Gale E. Christianson
(Penguin, 2000)

The Future of Ice: A Journey into Cold
Gretel Ehrlich
(Vintage, 2005)

The Weather Makers: How Man Is Changing the Climate and What It Means for Life on Earth
Tim Flannery
(Atlantic Monthly Press, 2006)

An Inconvenient Truth: The Planetary Emergency of Global Warming and What We Can Do About It
Al Gore
(Rodale Books, 2006)

Global Warming: The Complete Briefing
John Houghton
(Cambridge University Press, 2004)

Field Notes from a Catastrophe: Man, Nature, and Climate Change
Elizabeth Kolbert
(Bloomsbury USA, 2006)

The Winds of Change: Climate, Weather, and the Destruction of Civilizations
Eugene Linden
(Simon & Schuster, 2007)

End of the Earth: Voyages to Antarctica
Peter Matthiessen
(National Geographic, 2004)

The End of Nature
Bill McKibben
(Random House , 2006)

With Speed and Violence: Why Scientists Fear Tipping Points in Climate Change
Fred Pearce
(Beacon Press, 2007)

Hell und High Water: Global Warming — the Solution and the Politics—And What We Should Do
Joseph Romm
(William Morrow, 2006)

Safe Trip to Eden: 10 Steps to Save Planet Earth from the Global Warming Meltdown
by David Steinman
(Thunder's Mouth Press, 2006)

The Discovery of Global Warming
Spencer R. Weart
(Harvard University Press, 2004)

FILM AND VIDEO DOCUMENTARIES

An Inconvenient Truth
(Paramount Classics/Participant Productions, 2006)
A PowerPoint presentation on steroids, this filmed version of the former Vice President's lecture on climate change has succeeded in raising public awareness of global warming.

The Great Warming
(Karen Coshoff/Stonehaven Productions, 2006)
Comparing global warming to earlier disasters like the Black Death and the Great Depression, this film "should be required viewing by all," said the New York *Times.*

Too Hot Not to Handle
(HBO, 2006)
A guide to the consequences of global warming for the U.S., both current and projected.

What's Up with the Weather?
(PBS, joint production of *Nova* and *Frontline, 2006*)
A collaboration of two award-winning PBS series, this documentary investigates both the science and the politics of global warming.

Global Warming: The Signs and the Science
(PBS, 2005)
"It's not your grandfather's climate," warns this award-winning program, which aired only weeks after Hurricane Katrina devastated the U.S. Gulf Coast.

Strange Days on Planet Earth
(National Geographic, 2005)
A four-part series constructed as a high-tech, ecological detective story, with the fate of the planet hanging in the outcome.

Hot Planet, Cold Comfort
(PBS, 2005)
Part of the *Scientific American: Frontiers* series, this 60-min. segment focuses on how melting Arctic glaciers could potentially wreak planet-wide havoc.

Global Warming:
Science and Solutions
(National Center for Atmospheric Research, 2006)
Actor Erick Avari hosts this 120-min. look at the history of man's impact on his environment; experts offer pro-posed remedies to the current crisis.

GLOBAL WARMING FOR KIDS
Websites
U.S. Environmental Protection Agency
www.epa.gov/climatechange/kids/index.html

Pew Center on Global Climate Change
www.pewclimate.org/global-warming-basics/kidspage.cfm

Global Warming Kids
globalwarmingkids.net

Global Warming Challenge
www.globalwarmingchallenge.org

TIME FOR KIDS
www.timeforkids.com/TFK/specials/articles/0,6709,1113542,00.html

Books
This Is My Planet: The Kids' Guide to Global Warming
Jan Thornhill
(Maple Tree Press, 2007)

Why Are the Ice Caps Melting?:
The Dangers of Global Warming
(Let's Read and Find Out)
Anne Rockwell and Paul Meisei
(Collins, 2006)

The Down to Earth Guide to Global Warming
Laurie David and Cambria Gordon
(Orchard Books, 2007)

ON THE WEB
United Nations' Intergovernmental Panel on Climate Change
www.ipcc.ch/

Union of Concerned Scientists
www.ucsusa.org/global_warming

U.S. Environmental Protection Agency
www.epa.gov/climatechange/index.html

U.S. National Oceanic and Atmospheric Administration
lwf.ncdc.noaa.gov/oa/climate/global-warming.html

Pew Center on Climate Change
www.pewclimate.org/global-warming-basics

Sierra Club
www.sierraclub.org/globalwarming

Natural Resources Defense Council
www.nrdc.org/globalwarming/

National Wildlife Federation
www.nwf.org/globalwarming/

Columbia University's Educational Global Climate Modeling
www.edgcm.columbia.edu/

NASA Goddard Institute for Space Studies
www.giss.nasa.gov
(main site)
gcmd.gsfc.nasa.gov/index.html
(Global Change master directory)

Science Daily Global Warming Portal
www.sciencedaily.com/news/earth_climate/global_warming

New York *Times* Global Warming Portal
topics.nytimes.com/top/news/science/topics/globalwarming/index.html?inline=nyt-classifier

Live Science Global Warming Portal
www.livescience.com/globalwarming

Encyclopaedia Britannica
www.britannica.com/eb/article-9037044/global-warming

An Inconvenient Truth
www.climatecrisis.net

CONTRARIAN VIEWPOINTS
Though the vast majority of world scientists working in areas involving climate change agree on the broad outlines of the subject—global warming is real, it represents a worldwide crisis, and human activities are helping drive it—some people remain unconvinced, even if their ranks are dwindling. Here are some books, websites and a documentary program that reflect this point of view.

Books
The Politically Incorrect Guide to Global Warming (and Environmentalism)
Christopher C. Horner
2007, Regnery Publishing

The Skeptical Environmentalist: Measuring the Real State of the World
Bjorn Lomborg
2001, Cambridge University Press

Meltdown: The Predictable Distortion of Global Warming by Scientists, Politicians, and the Media
Patrick J. Michaels
2005, Cato Institute

Shattered Consensus: The True State of Global Warming
Patrick J. Michaels
2005, Cato Institute

Websites
Global Warming/Cooler Heads
www.globalwarming.org/

Global Warming Skeptics
www.globalwarmingskeptics.info

Documentary
The Great Global Warming Swindle
2007, WAGtv
This anti–*Inconvenient Truth* is the work of the U.K.'s most notorious contrarian, Martin Durkin (think Michael Moore, except British and right-wing), who offers a cynical look at the scientific evidence supporting global warming.

■ Postscript

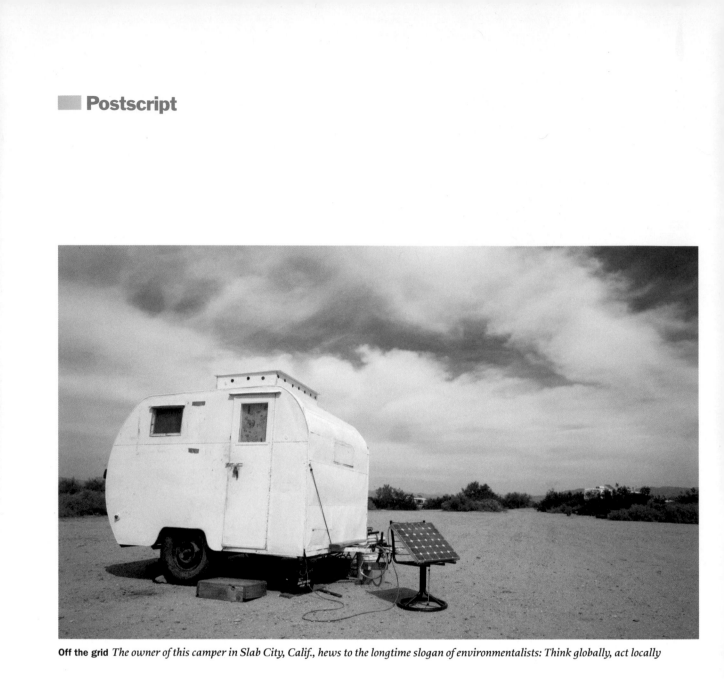

Off the grid *The owner of this camper in Slab City, Calif., hews to the longtime slogan of environmentalists: Think globally, act locally*